Genes, Germs and Medicine

The Life of Joshua Lederberg

Genes, Germs and Medicine

The Life of Joshua Lederberg

Jan Sapp

York University, Canada

World Scientific

NEW JERSEY • LONDON • SINGAPORE • BEIJING • SHANGHAI • HONG KONG • TAIPEI • CHENNAI • TOKYO

Published by

World Scientific Publishing Co. Pte. Ltd.
5 Toh Tuck Link, Singapore 596224
USA office: 27 Warren Street, Suite 401-402, Hackensack, NJ 07601
UK office: 57 Shelton Street, Covent Garden, London WC2H 9HE

Library of Congress Cataloging-in-Publication Data
Names: Sapp, Jan, author.
Title: Genes, germs and medicine : the life of Joshua Lederberg / Jan Sapp.
Description: Hackensack : World Scientific, 2021. | Includes index.
Identifiers: LCCN 2020056722 | ISBN 9789811225475 (hardcover) |
ISBN 9789811235986 (paperback) | ISBN 9789811225482 (ebook) |
ISBN 9789811225499 (ebook other)
Subjects: LCSH: Lederberg, Joshua--Health. | Bacterial genetics. |
Genetic engineering--Moral and ethical aspects. |
Artificial intelligence--Moral and ethical aspects. | Molecular biologists--Biography.
Classification: LCC QH434 .S27 2021 | DDC 579.3/135--dc23
LC record available at https://lccn.loc.gov/2020056722

British Library Cataloguing-in-Publication Data
A catalogue record for this book is available from the British Library.

For any available supplementary material, please visit
https://www.worldscientific.com/worldscibooks/10.1142/11971#t=suppl

Printed in Singapore

For Carole

Acknowledgments

It is a pleasure to thank the Richard Lounsbery Foundation of Washington for their support of this research. I probably would not have begun this project without the encouragement and enthusiasm of Rod Nichols and Jesse Ausubel. Ann Ganesan and David Kirsch offered helpful comments on an earlier draft of this book. Carole McKinnon has worked with me throughout this project and provided invaluable assistance with editing and advice on several drafts.

For their generous time in interviews, I am grateful to Jesse Ausubel, Walter Bodmer, Vint Cerf, Richard Danzig, James Darnell, Sidney Drell, Edward Feigenbaum, Anne Ganesan, Eugene Garfield, Richard Garwin, Cynthia Greenleaf, David Hamburg, Peggy Hamburg, Lee Herzenberg, Maren Imhoff, Robert Kahn, David Kirsch, Marguerite Lederberg, Rodney Nichols, Paul Marks, Matthew Meselson, Stephen Morse, Gus Nossal, William Perry, Thomas Rindfleisch, Vittorio Sgaramella, David Thaler, Mike Young, Mary Jane Zimmerman, and Harriet Zuckerman.

I would also like to thank archivists at the Cold Spring Harbor Laboratories, The Rockefeller Foundation, Stanford University, the University of Wisconsin, Madison, and the U.S. National Library of Medicine.

Preface

I first met Josh Lederberg in 1988. He had written to me in Melbourne after my first book *Beyond the Gene* was published, and I travelled to New York to meet with him while working on another project. He was president of the Rockefeller University then. Slightly disheveled, and with his sleeves rolled up, he and I argued about this and qualified about that as he climbed up and down the ladder to retrieve books from his floor-to-ceiling office library, always to be sure that we got things right.

We were like two kids in a sandbox. When I asked him how he could remain an intellectual while being president, he said that he was "a minimalist." After an hour or so, his secretary Mary Jane Zimmerman came to the door to say that he had a meeting. He quickly rolled his sleeves down, straightened his tie, put on his suit jacket, and just before two men in pin stripped suits came in, he turned to me and said, "Jan, can you come back in an hour or so."

Two years later, he invited me to come to the Rockefeller for a year as an Andrew Mellon Fellow. At that time, I was on sabbatical leave from the University of Melbourne and a visiting professor as Lynn Margulis' guest at the University of Massachusetts, Amherst. I was working on the history of research on symbiosis in evolution. By then, Josh had only recently stepped down as President of the Rockefeller and he was interested in that same topic as he had written a far-sighted paper on the importance of symbiosis in evolution forty years earlier.

He was away in Washington for a few days every week during that year working on national security issues and pandemic readiness, and sometimes in Geneva working for the World Heath Organization. But when he was in his office, I found that he always had time to talk and sometimes argue with me about my views on the history and sociology of science, or sometimes he would ask a question out of the blue. I later learned that he often did that. Once he walked up to me and asked how many genes there might be in the human genome, and when I said maybe 100,000, he asked *why* I thought that.

It would be a mistake to ever assume that he hadn't already read (or even himself written on) something that you thought was "new." I remember once, when I had first met him, I was naively "explaining" something to him about the role of symbiosis in the origin of cells when he stopped me suddenly and said, in an almost expressionless and matter-of-fact way, that he was "at least three steps ahead of me" in the argument I was making. I laughed. One time, when Lynn Margulis came to give a talk at the Rockefeller, I saw him fall asleep after 10 minutes or so, only to wake up at the end of her talk and ask a question on a central issue that she had addressed when he was asleep. When I asked him how he did that, he said, "It didn't take very long to know where she had to go."

I drafted *Evolution by Association* that year in Josh's lab, and we talked at great length about our microbial partners, how they have domesticated us, and how they would constitute an extended genome having had an important effect on our physiology and development. We were on email shortly after that, and he would send me references on various topics we had discussed. I remember him sending me a reference to a recently reported phenomenon confirming something that he'd said to me three years earlier, but which had been at odds with accepted wisdom. That reference came with one word: "see".

I visited Josh periodically in New York thereafter. In the mid-1990s, I went to talk with him about what I saw as a newly emerging evolutionary and ecological viewpoint with regard to infectious disease outbreaks: that rather than simply waging war against them, we would have to understand them from an evolutionary and ecological perspective. What impressed me and probably everyone else about Josh was not just his breadth and depth of knowledge or how much he read, but the speed of his thinking and the rigor of his argumentation.

Josh invited me to a meeting on the future of genetics and biomedical research at the National Institutes of Health, held in the dark shadow of the fallen Trade Towers in November 2001. The discussion turned to how narrow and rigid science education had become, and how he himself would not be there that day had the system been so inflexible when he was a student. I didn't know what he meant by that until I began writing this book.

I gave a talk on Josh's views about symbiosis and the superorganism during his 80th birthday celebration at the Rockefeller, and then saw him for the last time two years later in the summer of 2007. He was ill and in pain as he hobbled into his living room using a walker. Yet, he seemed to forget all that and light right up when we discussed microbial evolution.

This book has been a long time in the making. Josh had first asked me to write his biography in the late 1990s and proposed some ideas about how he and I might collaborate on it, but that never happened. This book, however, is also a collaboration of sorts because it could not have been written without the immense amount of information he left behind in his archives. In researching this book, it became clear to me just how little I, and everyone else it seemed, knew about his extraordinary career. Everyone who knew him perhaps saw only one part of the elephant in a dark room. Herein, I have sought to shed light on the whole person.

Contents

Chapter 1

The Polymath

"Well, I've kind of a curse. I'm interested in everything."
Joshua Lederberg[1]

Joshua Lederberg's contemporaries knew him to be one of the greatest scientists of the twentieth century, a creative mastermind, "a universal genius" — brilliant biologist, mathematician, computer scientist, statesman, philosopher, historian, and writer.[2] It's been said that the word polymath was invented for him. His "home field" was genetics, for which he was awarded the Nobel Prize, but he also made fundamental contributions to immunology, evolutionary theory, the search for extraterrestrial life, molecular medicine, biological engineering, artificial intelligence, and the study of emergent viral diseases.

He conceived of his Nobel Prize research when he was just 20 years old and began carrying it out while on temporary leave from medical school in 1946. In 1958, he became the second youngest Nobel Prize winner for medicine in history; only Frederick Banting was younger (by a year) when he was awarded the Prize in 1923 for the discovery of insulin. Lederberg's Nobel Prize was not for discovering a medical treatment or cure, but it laid the foundation for the genetic engineering of bacteria from which many hormones and drugs, including insulin, are produced today.

Although Lederberg was part of the golden age of molecular biology when the genetic code was deciphered, and genetic engineering was envisaged, his research career began when DNA was not yet known to be the basis of heredity and when bacteria were thought to lack genes. He opened an entire new field of bacterial genetics upon which much of molecular biology is based when he corrected the latter misconception and showed that bacteria possess genes like other organisms.

His colleagues saw him as a visionary and it's easy to see why. Today, he is perhaps best known for his early warnings about plagues to come.[3] In the 1960s and 1970s, when the "war" against microbial disease was falsely declared won, Lederberg sounded the alarm about potential global pandemics and argued that science was ill equipped to prevent or even mitigate them. He repeatedly called for international cooperation because he foresaw the need for systematic vigilance against emerging diseases. He continued to press for preparedness in the 1990s when government agencies were sluggish and disinterested in taking action. At first perceived as an alarmist, he was later hailed as a prophet. A deep evolutionary understanding underlay his thinking about virus pandemics. Humans, he repeatedly argued, were both inside and outside of nature and "nature" itself is far from benign. The future of humanity and microbes would be a battle of "our wits versus their genes."

Lederberg also foresaw the biotechnology that would emerge from molecular biology. Before the genetic code was even cracked, he explained how genes could be modified specifically through the insertion of a synthesized DNA segment — a process he called "directed mutagenesis." That field didn't emerge until decades later. He was one of the first to see how recombinant DNA research might be achieved — by inserting genes into bacteria to make medical products — and he urged biomedical companies to develop such products when they were reluctant to do so. He was among the first to advance the use of computers in medicine and for scientific communication. Sixty years ago, he predicted that DNA might store information much like a computer does — a technology that is only now becoming a reality. He also forecasted the general use of e-mail and other electronic forms of communication decades before they were a reality for the rest of us.

Lederberg wrote extensively about the social and ethical ramifications of the biotechnology he foresaw. He had a weekly column in *The Washington Post*, which ran for five years during the turbulent 1960s and early 1970s. He consistently provoked readers with his unapologetic views on many of the controversial issues then and now: abortion, race and civil rights, environmental risks, biological engineering, as well as warnings of the risks of viral pandemics in an unprepared world. His humanist rationalist viewpoints remained unassailable and as steadfast as his belief in the progress of biomedical science and its importance for human welfare.

Lederberg had an unusual talent for spotting new lines of scientific inquiry and getting new research programs started. He had an enthusiasm for interweaving disciplines, a gift for seeing how methods and theories in one field might apply to another, and a knack for forging unexpected collaborations. His community of collaborators included chemists, psychiatrists, engineers, computer scientists, sociologists, historians, poets, and government leaders.

At a time when genetics was essentially ignored in medical education, he founded medical genetics departments in the United States, first at the University of Wisconsin and then at Stanford's medical school where his eclectic interests expanded. He not only advised NASA on the field that he called "exobiology" or the search for extraterrestrial life, but also helped design the instruments needed to test for life on Mars. He was also the first to propose that organic molecules formed in space may have stimulated the origins of life on earth; data supporting this proposition manifested decades later. The technologies he helped to develop for NASA would also have major ramifications for various aspects of medical science. Collaborating with computer scientists, he created and developed the first artificial intelligence systems that could be applied to science and medicine.

During the 1960s and 1970s when many of his colleagues abhorred the idea of offering scientific advice on matters of national and international security, Lederberg became committed to it. He was an advisor to nine Presidential administrations from Eisenhower to G. W. Bush and was engaged in national and international policy in biomedicine, biowarfare prevention, and public health. Like the physicists who loathed the atomic bomb they had made, he worked against the development of germ warfare using methods he had helped to create. A witness of the Cold War and a child of the Great Depression, he was an internationalist in policy. He was instrumental in establishing an international treaty banning biological weapons, and he led negotiations on compliance with Soviet counterparts.

A scientist's individual style is as important as that of any artist. Lederberg's was formed early in life when, as a child of poor orthodox Jewish immigrants growing up in New York City, he rejected religion as a vocation in favor of science, and then chose to follow a humanist rationalism wherever that might lead. He was a brilliant prodigy with a forbidding personality. He threw himself into science, history, philosophy, and anything else he could find to read about as a schoolboy, even etymological tomes. Throughout his life he sprinkled the dictionary of biology with neologisms which have become part of the common parlance of biologists.

"Microbiome," the last word he coined, is perhaps one of the best known outside of science.

Dedicated teachers and professors mentored him through high school and university. He was kept out of the war and entered medical school in the navy reserve. In a world in which talent overcame poverty and the anti-Semitism he sometimes faced, he dropped out of medical school to pursue his research as a 22-year-old professor with a little laboratory in Madison, Wisconsin. In the decade following World War II, he and a small band of young collaborators published groundbreaking paper after paper, thus constructing much of the field of bacterial genetics in methods, concepts, and vocabulary.

Lederberg's career path followed the principal scientific and technological developments in biomedicine since the Second World War. It takes us on a tour through the social, ethical and political aspects of those developments — from the intimate workings of the laboratory to the politics of the Nobel Prize and the world of national and international decision-making on some of the most critical issues of our time.

Lederberg's road was far from smooth. His diary notes tell how his success as a scientist was frustrated by a troubled marriage with his first wife, Esther, herself an accomplished biologist and collaborator in much of his early genetic work. His archives reveal the workings underlying his being awarded the Nobel Prize in 1958 and his disinclination to accept it. He suffered a deep depression in the years following the award but was able to reinvent himself after he moved from Wisconsin to a faculty of medicine like no other at Stanford. Assembling a small albeit eclectic department of genetics, he expanded his research foci and forged new collaborations for himself and for his young faculty members.

After nearly two decades at Stanford, he transformed himself yet again and returned to New York City to become president of one of the most important biomedical research institutions in the country, the Rockefeller University. His government advisory work increased in intensity and focused on germ warfare prevention and preparedness for plagues to come.

The Lederberg story is a journey through the biomedical sciences and their salient societal effects in war and peace. But it is also a story of how the son of poor immigrants was nurtured by a social system that Americanized him and kept him out of the Second World War, how he achieved success and fame, and how he paid back that debt to his country.

Chapter 2

Heretic

What I Would Like to Be.
I would like to be a scientist of mathematics like Einstein. I would study science and discover a few theories in science.

Joshua Solomon Lederberg, June 20, 1932, Age 7[1]

Joshua was the first of three sons born to Esther (née Goldenbaum) and Zvi Lederberg, an orthodox rabbi, both natives of Yafo Palestine. There was a long line of religious Polish Jews on his father's side who had emigrated to the Holy-Land.[2]

Fig. 2.1. Zvi and Esther Lederberg at their wedding, 1924, The Joshua Lederberg Papers, Profiles in Science, The National Library of Medicine.

The tradition of rabbinical scholarship was even stronger on his mother's side. Hers was a Hassidic rabbinical family from the town of Safed in the north of present day Israel. Her father, Rabbi Aharon Goldenbaum, was a member of Israel's ultraorthodox Gur Hassidic sect. When the Turks began hunting for him before the First World War, he fled to America where he was appointed Rabbi of the Roumanian-American Synagogue in Brooklyn.[3]

Joshua's mother had very little formal education, which was typical for girls from orthodox families, but his father was well educated in traditional seminary fashion. A brilliant scholar, he had been sent to the United States for studies as a teenager and received the equivalent of a green card at that time, so he was able to immigrate. But first he went back to Palestine in 1924 to claim his bride in an arranged marriage.[4] The Lederbergs came to the United States the following year.[5]

Joshua was born May 23, 1925 in Montclair, New Jersey, and six months later the family moved to Washington Heights in Manhattan where his father took up a position as rabbi in a small orthodox synagogue.[6] His younger brother, Seymour, was born on October 30, 1928. Seymour's presence introduced young Joshua to "executive responsibilities" while his parents were fully occupied "bringing home what must not be called the bacon."[7] Josh treated Seymour as his student and mentored him, and Seymour would eventually become Professor of Microbiology and Genetics at Brown University. Their much younger brother, Dov, born in 1941, became a filmmaker and photographer, and in 1965, a Chabad Chassid, an Orthodox Jewish mystic.

Virtually penniless when they arrived in the USA, the family struggled to make ends meet during the Great Depression. Rabbi Lederberg was like the parson in a small town — low in cash but high in respect. They were often helped by shopkeepers, with gifts of meat from the butcher and bread from the baker, but there was little money to pay the rabbi at the synagogue. It was the five-dollar fee or fifteen-dollar fee for officiating at weddings and funerals that kept the family going. It was Joshua's job to guard the telephone and make sure that no calls requesting the rabbi to officiate at a special ceremony were ever missed.[8] His father fell ill in 1932 with progressive ulcerative colitis, and by the end of the decade, he was only able to work part-time. His mother vigorous and intelligent, learned English and worked various jobs outside the home, including teaching in a Hebrew school and catering to help keep the family together.[9]

Joshua's parents returned to Israel in the early 1960s. The walls in their Tel Aviv apartment were festooned with copies of the degrees, diplomas, certificates, and awards received by their American sons.[10] In 1967, his mother recalled with enthusiastic pride a moment from seven years earlier: she was chairing a meeting for a Zionist women's organization in New York when Senator Jack Kennedy gave a lecture. Kennedy knew about Josh who had recently been awarded the Nobel Prize at the young age of 33:

When he saw me wearing a pin with the name Lederberg, he came straight over, shook my hand, and asked, "Are you perhaps a relative of Professor Lederberg?" I replied, "That's my son! And if he wasn't too young to receive The Nobel Prize, you are not too young to be President." Kennedy laughed, and I added, "If you are elected President, please send me an invitation to the inauguration." And what do you think? That he didn't send? He sent. But at the time there was a heavy snowfall and I couldn't get to Washington. But Joshua participated in the ceremony.[11]

Shades of Spinoza

As the eldest son, it was expected that Joshua would become a rabbi like his father as well as his father before him. Though he was devoted to his family, Joshua had other ambitions and struck out in the direction of science. "I seem to have been born a scientist as far as I can tell," he commented later in life. "Learning and a quest for truth were paramount in my family background but there were no scientists, no physicians, no academic people anywhere in my family ambiance."[12] Much of his interest in science came from his Americanization. He was "the melting pot American."

His keen interest in science made him something of a heretic. Like the great Dutch philosopher Baruch Spinoza, who had helped to lay the groundwork for the 18th Century Enlightenment, young Joshua challenged his father on the basic tenets of Orthodox Judaism. He identified with the fundamental, ethical, and philosophical traditions of Judaism but not the practices, which he thought were simply medieval. He was eager to be secularized, to be in the American tradition.[13]

There was tension between him and his father but a lot of mutual sympathy too. His father eventually gave in and said, "Joshua, I guess there are many ways to follow the Torah and if you want to seek truth through science, that's alright too."[14] Once his father reconciled with the idea that Joshua was not going to be a rabbi and would be straying from routine observance, they were able to discuss a lot of things from their respective positions with mutual respect.[15] His mother seemed to be more amenable to her son's interest in science as a vocation. After all, it was she who had proudly saved a composition of his from second grade in which he said he wanted to be a scientist like Einstein.

A heretic to the faith, he rubbed against its rituals. Libraries were forbidden on the Sabbath, but Saturdays were the best days to go to the library. He would walk a mile so that none of his father's congregants would see him, and then get on the subway to go downtown to the New York Public Library. That library was a critical element of his early education and Americanization.[16] So too were the public schools in New York, which were well calculated to be "instruments of Americanization." His teachers nurtured his interest in science, but he learned more from books than from classes. He wanted to read everything that science had to offer and everything outside of science that he could assimilate — history, politics, philosophy, and some literature.[17]

The family residence was on Haven Avenue in Washington Heights, and the local public school he attended was right on the northern border of Harlem. Many African American kids attended the school, but Joshua noticed no great discrimination then and a minimum of racial strife. There were pockets of Irish kids in this mostly Jewish community who listened to the priest talking about "Christ-killers," but the kids knew where the boundaries were drawn and the allowed zones for walking to and from school. Any fool who strayed from those routes would be beaten up.[18]

There were standardized IQ tests of school kids in those days and the results were even announced publicly. Lederberg was said to have had the highest score of all in the eastern United States.[19] He fell right into the melting pot and embraced the view of America "as the land of opportunity," a well-founded view indeed as it turned out.[20]

Like many gifted children, Joshua was very lonely and isolated from his peers in grade school. His schoolmates knew he was different; they were kind of baffled but tolerant of him.[21] Though he was socially awkward, he was unabashed intellectually and devoted teachers encouraged him — even when he sometimes teased them with questions they could not always answer or mathematical riddles they could not solve.[22] Those antics eventually led to a truce whereby he was permitted to sit at the back of the room and study what he wanted.[23] His interest in science continued to grow and he soon switched his focus from mathematics to medicine, chemistry, and biology.

At the age of ten, he read in *The New York Times* that Wendell Stanley had purified a virus in the form of needle-shaped crystals with the chemical properties of proteins. That discovery prompted the question, "what is life?" At eleven, he attended a lecture by renowned cell biologist Robert Chambers and he was bold enough to stand up to ask if the spindle fibers in cells were real or artifacts of the microscopic staining and fixing procedures — a central unresolved issue of the day.[24] One of his teachers, Mrs. Fanny Rippere, wrote about the 11-year-old: "Early in 1937, I had a most unusual pupil whom I still remember vividly. I can still remember how he prepared a paper on the classification of Protozoa using a graduate text for a reference."[25]

The only books in his home that were of use for secular learning were the dictionary and the encyclopedia. He made good use of both, reading them from cover to cover. He studied words and their origins, and his vocabulary was as astounding as his memory. He was the spelling champion for his school, which he represented at Radio City Music Hall. With an advertisement coupon, he bought a dictionary that had some etymology in it and taught himself the Greek and Latin roots.[26] He would use that knowledge later on when coining several terms now commonplace in the biological lexicon.

Albert Einstein and famed biochemist, Chaim Weizmann, who would be elected the first President of Israel, were his cultural heroes. The ambition they inspired was strengthened by popular culture that romanticized the medically oriented scientist. He read Paul De Kruif's best seller, *Microbe Hunters* (1926), and novels such as *Arrowsmith* (1925) and *The Life of Louis Pasteur* (1928). These kinds of books turned a whole generation toward a career in medical research. There were inspiring movies as well. *The Symphony of Six Million* (1932) told of the rise of a Jewish physician from humble roots to the top of his profession. *The Magic Bullet* (1940) was about the life of Paul Ehrlich, whose research on immunity earned him the Nobel Prize in 1908. These biopics lionized men who pursued and valued medical science above all else while often neglecting their personal lives — marriage and family.[27] As a hyper-aware youth, Joshua was deeply affected by them.

Fig. 2.2. Joshua Lederberg at the microtome, The American Institute of Science, 1941. The Joshua Lederberg Papers, Profiles in Science, US National Library of Medicine

Fig. 2.3. Lederberg giving a lecture, biology symposium, American Institute August 1941, The Joshua Lederberg Papers.

As for the study of science itself, *The Science of Life* (1934) by H. G. Wells, Julian Huxley, and Wells' son, George, gave Lederberg a perspective on biology and the place of humans in the cosmos. Medical books were not readily available to him then, but he was able to read about cell structure, microbiology, and immunology in high school. He received Meyer Bodansky's textbook, *An Introduction to Physiological Chemistry*, as a present on his Bar Mitzvah in 1938.[28] His second treasured possession was E. B. Wilson's masterwork, *The Cell in Development and Heredity* (1925), which was given to him on his 16th birthday.

Stuyvesant

New York provided other special opportunities for young Lederberg. He passed the admission exams for Stuyvesant High School, which specialized in science and mathematics and collected some of the best students from all five boroughs. Lederberg had quite a commute from Washington Heights downtown to the school on East 14th Street, but he used the time get his homework done. The most important thing that Stuyvesant offered him was a peer group of youngsters who had similar interests in science and who, he said, "could sharpen, in some competitive way, excitement about pursuing various things."[29]

Stuyvesant was an all-boys school then, and when he found three or four boys that he was comfortable with, he no longer felt the isolation. About two thirds of the students were Jewish and, like him, second generation immigrants. He joined all the science clubs — astronomy, chemistry, biology — and worked actively in them, spending quite a bit of time doing so after school. He became president of the Biology Club and the Medical Society, vice-president of the Bunsen Chemistry Society, and president of the Jr. Chemistry Society. He also gained entrance to the Cooper Union Library a few blocks away from Stuyvesant, where he was able to read scientific research papers and *The New York Times*.

Joshua was also offered space for a year at the American Institute of Science Laboratory (a forerunner of the Westinghouse Science Talent Search). Located in the shadow of the Empire State Building 15 minutes from school and 45 minutes from his home, it offered facilities to conduct research after school hours and on weekends. He studied the chemical basis of staining and fixing cells.[30] He was just 15 years old when he graduated from high school in February 1941. He worked in the downtown laboratory during the spring, ready to enter university in September.

But not everything was biology and chemistry. To Joshua, the world was full of interesting puzzles. One experiment he was involved in yielded very mystifying and frightening results that still baffled him, as he related it, fifty years later. When he was about 14 years old, he and a friend became interested in hypnosis, read up on it, and thought they'd give it a try on a boy they knew from junior high school.[31] He was not an academically minded boy by any means. The boy felt guilty about masturbation and they offered to help him with that if he'd be their subject. He was very willing, very suggestible. They got him to the point where the code, "Oom, oom, sleep!" would make him go right under. He was conditioned.

Then they got the shock of their lives. First, they asked him to think back to when he was an infant, and he gave appropriate responses. Then they asked him what he was before that, and what he was before that as Lederberg recalled:

> 'Were you ever reincarnated?' He said, 'Of course!' We said, 'Well, let's go back. What are you now?' Before long he was a scribe in Egypt, and we were asking him to describe his environment. Here was a kid who was barely literate, and he started writing out hieroglyphics... We got a couple of pages of this kind of stuff, and we were trying to figure out if we could translate it, if we could figure it out.

They were pretty scared when he first started writing hieroglyphics. They just didn't know what genie they had let out of the bottle. Where did all this come from? There was no way this boy had ever even heard of a hieroglyph, and when he was awake he said he had no knowledge of any of it.

They went to see one of the Egyptologists at the Metropolitan Museum of Art and asked him if he could identify the writing. After looking at it for a while, he said that it looked like a popularization of the mid-nineteenth century work of the scholar and philologist Jean François Champollion. He could identify it as such by certain mistakes Champollion typically made in his renderings. How that boy ever picked that up remained a mystery.

But the incident taught Lederberg "to just never underestimate anybody's intellectual potential; it can be overlain with all kinds of things, and if you only get to the root of it, you can get all kinds of fantastic productions. I have no idea what's happened since, and I have no idea whether we 'cured him of his habit.' I'm not even sure what our view on the matter was, but anyhow there you are."[32] Lederberg never seemed to lose this open-mindedness when exploring ideas, no matter how unorthodox.

Chapter 3 In Navy Uniform

> I earnestly hope that Mr Lederberg's case will be given very thoughtful consideration so that a man of such rare ability will be most effectively utilized for the war effort and yet still be with us when the war is won.
>
> Frank Stodola to US Navy Reserve V-12 Program
> December 28, 1942[1]

Lederberg considered several universities for the fall of 1941. He applied to Cornell in Ithaca but failed to receive a needed scholarship for tuition, room, and board because, he suspected, he was Jewish. "Cornell was quite discriminatory," he later recalled, "a farm boy could get into the program I had in mind, I couldn't." Very few New York City students could.[2] He considered City College of New York, which was downtown and he could stay at home and commute. A lot of brilliant students went there. But it was crowded and had limited laboratory facilities, so he considered it a last resort.

He aimed for Columbia University on the Upper West Side. He knew that Thomas Hunt Morgan and his *Drosophila* school of genetics had developed the chromosome theory of heredity there. Morgan was the first non-medical person to be awarded the Nobel Prize in Medicine in 1933. Fortunately, Lederberg was awarded a tuition scholarship of $400 from the Haydon Trust. "I think that being able to go there was the luckiest thing that ever happened to me," he commented decades later.[3] Indeed it was, because five years after entering the university, he would make a discovery that would eventually change genetics and much of biology.

Nothing was normal about his Columbia experience. As a zoology major, he passed some courses without having to attend them because of his own personal study, and he was allowed to enroll in several postgraduate courses even though he was still very much an undergraduate. Professors got to know him for his extraordinary memory, quick mind, and breadth and depth of knowledge that often exceeded their own.

For the first two years, Lederberg commuted to the university from his parents' home in Washington Heights, three miles north.[4] Though he was an undergraduate, he tagged along with graduate students and even postdocs — people who were 5 to 10 years older than him. They "would look askance when they first met me," he recalled, "but after they got used to my manners would accept me."[5]

Professors who allowed him to take courses without prerequisites did so at their own risk. The renowned cell biologist, Gordon Whaley, recalled an incident when, as an untenured instructor at Columbia, he was rebuked by his superiors for his assessment of Lederberg's term paper. He thought it was "superb" and displayed "independent and adventurous thinking." But it was quite unlike any other paper he'd ever received, so he asked one of the senior members of his department to read it. The senior professor thought it was awful and advised Whaley to give it an "F". After two sleepless nights, he decided to give it an A+. "As an upshot," Whaley recalled decades later, "I got a scorching letter from the professor and a notation that when the question of tenure came up he intended to vote against it. Fortunately, it never came up. I don't think I taught Joshua anything, but he taught me some things."[6]

The United States entered World War II three months after Lederberg enrolled at Columbia. He had long been attuned to the war as the horror unfolded. He knew well of the persecution of the

German Jews and the flight of intellectuals, including Einstein. He followed events closely — from the occupation of Austria, the Nazi-Soviet pact, and partition of Poland, to the fall of France and the Battle of Britain. When the Japanese bombed Pearl Harbor on December 7, 1941, he was only 16, too young for conscription; it would be another two years before he would be old enough to be drafted.

Lederberg applied to the Navy Reserve's V-12 program in December 1942. The V-12 was established to train officers to meet the demands of the War. The Navy supported the student's education in return for later service. Some 125,000 students enrolled in colleges and universities in the United States through the program. Entering the V-12 program also meant that Lederberg would be protected from the draft.

Frank Stodola, a professor of organic chemistry, wrote a remarkable letter supporting his application.[7] Lederberg passed first year chemistry through an "achievement exam on the basis of personal self-instruction" and then took Stodola's chemistry course in the summer of 1942."[8] Stodola explained to the Navy Reserve that he had never met anyone like him before:

> I can start off by saying, that this young man is the most brilliant student I have encountered in my nineteen years of association with various universities. He not only has an amazing memory, but also has the rare gift of being able to think through the most difficult problems. He seems well versed in almost any branch of learning, especially science. His particular interest is biology, and for that reason he should be made use of in the medical training program. And what is more, he should be used as an instructor and not be forced to waste his time as a student.[9]

Lederberg took the examinations and entered the US Navy Reserve as Ensign and premedical student. The program was designed to compress premedical training to about 18 months of instruction and the four-year MD curriculum into three calendar years. There were no summer vacations. Students were paid $50 per month and were required to wear service uniforms and engage in physical training.

Lederberg was given residency in the dormitory on the Columbia College campus. Paul Marks, who was his roommate for two years, recalled how completely uninhibited Lederberg was academically, ready to stand up and interrupt professors' lectures with probing questions.[10] That is, if he bothered to attend class. If

Fig. 3.1. Lederberg in Navy Reserve Uniform, 1943, The Joshua Lederberg Papers.

the class was in the morning, he preferred to remain in bed and ask Marks for his lecture notes: "He'd read them over and do better than I did," Marks recalled.[a]

The V-12 pre-med program was designed so that students could be assigned to active duty for several months of the year. Those who had prior laboratory experience were put into the clinical labs. Lederberg ended up working in the pathology laboratory at St Alban's Naval Hospital on Long Island for three or four months every year over the course of his pre-medical training.[11] His experience there imbued him with a sense of life-long patriotic gratitude and duty.

St Alban's was as close as Lederberg got to the war.[12] Once in a while he would take care of patients, and sometimes he was on morgue watch. When a patient died during the night, his job was to remove the remains and help set the body up for autopsy. But his main job was in the laboratory. The Third Marine Division was at the hospital before and after tours of duty. Their attitude toward him and the other students in the V-12 program varied: Sailors who were about to be shipped out were resentful of them, but those who returned told them that they were lucky not to have to face combat.[13] Lederberg had friends who were drafted; several were killed.[14]

In June 1944, Lederberg was awarded a BA with "Honors in Biology," ready to begin medical school in the fall. The day after graduating with his Bachelor's degree, he was seconded for a third and last stint at St Alban's Naval Hospital. This time, he was busy doing stool examinations for hookworm and amoebic dysentery, and blood smears for malarial diagnosis for marines

---

[a] Their careers brought them in close proximity in the late 1970s when Lederberg was president of the Rockefeller University in New York and Marks was president of the Sloan Kettering Cancer Center adjacent to it (see Chapter 27).

returning from the Guadalcanal Campaign in the Solomon Islands. About two thirds had malaria; some had other intestinal parasites. The work was interesting to Joshua and he did it cheerfully, staining and examining the tiny chromosomes of the malarial parasite *Plasmodium vivax*. That work would be important in shaping his outlook on microbial genetics. Observing the stages of the malaria parasite alerted him to the possibility that there might even be sexual stages in the smallest of microorganisms, including bacteria.[15]

However, Lederberg's experience in medical school was less than thrilling. It wasn't anything like what he had hoped it might be. Barely a month into the program, he showed signs of discontent and complained about the tedious rote learning, dull courses, and lack of exposure to clinical aspects. "Medical school may be an even more insidious threat to one's pure scientific integrity because for pedagogic reasons the clinical facet is not exposed until very much later," he wrote to his former biochemistry professor Frank Stodola in November 1944.[16]

Lederberg was even more disheartened with medical school by the following summer as he wrote in a letter to his close friend from Stuyvesant High School, Jack Edman. He complained about sitting in stuffy dark rooms "while some old geezer talks on and on and on about measles or tuberculosis." He complained that medicine was not only "spoon fed", but was not much of a science at all as far as he could tell, because rather than being based on research to aid many patients, it focused on the singular patient: "Cure the patient, and as a result, probably less patients are cured in the long run. I don't like it, and probably will not practice very long after I get out."[17]

Hyperaware of the irreparable rupture with his friends caused by his sitting out the war as a V-12 student, he continued in his letter to

Edman: "People are melting out of my world, which is the way I have always pretended that I wanted it to be," he said. "I haven't made any new friends, since you guys, and I know damn well that even that can never be the same again . . . That's my price for having sat out the war."

Lederberg explained to Edman that he belonged in the laboratory conducting fundamental research. "The laboratory is more than just a dull place where you wash test tubes. There, and not on the dance floor, drill field, or battleground, I'm at my best." Indeed, he had actually gotten a taste of laboratory research at Columbia before he entered medical school, and that as much as anything else spoiled him for medicine.

A year or so after he had arrived at Columbia College as a zoology major, Lederberg met a young and very bright assistant professor, Francis Ryan. Ryan became his most important mentor and introduced him to the emerging field of microbial genetics. The new methods for their genetic study had just been worked out, and they were revolutionary. Ryan had recently learned them as a postdoctoral fellow at Stanford University.

The Genetics of Microbes

During the first four decades of the twentieth century, genetics was based on the study of corn, mice, and especially the fruit fly, *Drosophila*. These were the organisms geneticists used to map the positions of genes along chromosomes and identify visible traits, or phenotypes, that genes affect. But no one knew what genes were or how they affected those traits. There were speculations, though, that genes might operate by determining the nature of protein enzymes which control biochemical reactions.

To study how genes affect biochemical functions, geneticists needed to use microorganisms whose sex life and nutrition could be meticulously controlled. And they needed to collaborate with biochemists. Biochemists would isolate mutants that were unable to grow, or that grew poorly, on a well-defined growth medium, and geneticists would then conduct the crosses to identify the genes associated with the metabolic functions underlying those deficiencies.

These new methods were devised at Stanford by the now famous team of George Beadle and Edward Tatum. They began working together on a bread mold called *Neurospora crassa* in 1941, the year Lederberg entered university. He would later share the Nobel Prize with them for his work in establishing bacterial genetics. Beadle had been a corn geneticist and Tatum was a biochemist before they teamed up.

The idea of using *Neurospora* for genetics study came to Beadle when he heard Tatum give a biochemistry lecture in which he described the nutritional requirements of yeast and fungi.[18] It then occurred to Beadle that one should be able to select for mutants that were unable to synthesize known metabolites, such as certain vitamins and amino acids. A mutant incapable of making a certain vitamin, for example, could be grown when that vitamin was added to the culture medium, and would be identified as such when grown in media lacking that vitamin. *Neurospora* seemed to be ideal; its growth requirements had already been worked out by biochemists.

So in the spring of 1941, Beadle and Tatum began seeking *Neurospora* mutants by X-raying them or exposing them to ultraviolet light. In a key paper that year, they articulated what came to be standard methodology in the emerging field of biochemical genetics.[19] The Beadle and Tatum team would grow into the *Neurospora* school with a talented group of researchers.

The *Neurospora* work proved to be of great importance in the drug industry and in biochemistry, providing a powerful tool for

Fig. 3.1. Francis Ryan, circa 1950, The Joshua Lederberg Papers.

the analysis of the pathways through which vitamins, amino acids, and other compounds are synthesized. Despite the contingencies of the Second World War, a number of graduate students and post-doctoral fellows hurried to Stanford to learn the new *Neurospora* techniques. Francis Ryan was one of the first to do so.

Ryan had no microbiology background when he went to Stanford as a post-doctoral fellow in 1941. He had just completed his PhD at Columbia on a study of the embryological development of frogs, and he had gone to the west coast with the intention of continuing embryological research. But his plans changed when Beadle and Tatum reported their first findings on biochemical mutants in *Neurospora*. He implored them to accept him into their lab.

Lederberg did the same when Ryan returned to Columbia the following year as an assistant professor and brought *Neurospora* with him.[20] Lederberg knew of the work of Beadle and Tatum, and he had heard people speak highly of Ryan. He began as Ryan's dishwasher and assistant, but before long he was allowed to do more things around the lab, including developing ideas about what experiments to perform.[21]

Ryan was nine years Lederberg's senior, and although he was only a year out of his PhD, he had a paternalistic, nurturing attitude. He was a talented teacher who favored the Socratic method. Lederberg challenged him with intellectual sparring that they both enjoyed; they bonded. Ryan taught Lederberg how to conduct himself in the lab, sharpen his adolescent thinking, and confront scientific issues in a professional way. He learned how to do experiments, keep his notebooks, record his results, and become more focused.[22] Ryan was also a remarkably noncompetitive scientist and quickly realized, like others, that Lederberg's intellectual talent was matched by ambition.

There was a group of four or five people working in Ryan's laboratory during the War. He had a research contract from the military to study the nutrition of some wound-infecting bacteria and develop therapeutic methods to manage those infections.[23] Several people worked on that project.

Lederberg worked with Ryan on *Neurospora* and he taught Lederberg what Beadle and Tatum had taught him. But their interests were different — Ryan was more of a physiologist than a geneticist and was mainly focused on the biochemical substances that regulate the growth of *Neurospora*. Lederberg was interested in the chemical basis of the gene itself and proposed that they conduct some experiments on that.

No one was sure what the gene was in the 1940s. DNA was in the cell nucleus, but proteins were also components of chromosomes, and at that time they seemed to be the more likely candidates for the gene because they were known to be complex and diverse. All the available chemical data on DNA then seemed to indicate that it did not possess a high enough degree of complexity or variability to accommodate the enormous diversity of life. Chemists of the 1930s had concluded that DNA was a monotonously uniform macromolecule, like starch for example, and they thought DNA did not vary much between organisms.[24]

The first experimental evidence suggesting that DNA might be the stuff of genes came not from chemists but from studies of the virulence of two different strains of a pneumonia-causing bacterium called pneumococcus. In 1928, a London microbiologist, Frederick Griffith, reported that a nonvirulent strain of pneumococcus could be transformed into a virulent strain when mixed with a heat-killed virulent strain.[25] What was this transforming principle? Griffith was killed in his apartment during a German air raid on London in 1941. But Oswald Avery and his collaborators at The Rockefeller Institute for Medical Research in New York attempted to identify the transforming substance. They exposed the nonvirulent strain to purified protein, RNA, sugar, and DNA extracted from the virulent strain. Only the bacteria exposed to purified DNA were transformed. They reasoned, therefore, that the nonvirulent form had acquired DNA from the virulent strain. They reported their results in a famous paper in 1944.[26]

Young Lederberg took the pneumococcus results as a strong indication that DNA was the stuff of genes.[27] The precocious nineteen year-old wrote about it in his diary on January 20, 1945:

> I had all evening to myself, and particularly the excruciating pleasure of reading Avery '43 [actually 1944] on the deoxyribose of nucleic acid responsible for type transformation in Pneumococcus. Terrific and unlimited in its implications... Direct demonstration of the multiplication of transforming factor... Viruses are gene-type compounds.[28]

Lederberg knew though that the evidence that the transforming principle was actually the gene, and that it was DNA was not definitive. There were two alternative interpretations: the non-virulent form could have been modified somehow by a "specific mutagen," or else the purified DNA was not so pure. It was possible that a small amount of protein, as much as 1 or 2 percent, was present in their preparations.[29]

First, Lederberg wanted to know if transformations occurred in other microbes that were known to have genetic systems. So, in the spring of 1945, he suggested to Ryan that they try to see if they could demonstrate such transformations in *Neurospora*. In the spring of 1945, he suggested for Ryan that they should see if it occurred in *Neurospora*. They began those experiments in June, but no transformations were revealed. (In fact, no study has shown that purified DNA can be incorporated into the chromosomes of cells, other than in bacteria).

Their failure with *Neurospora* was a blessing for Lederberg because it led him to bacterial genetics. At that time, it was not even certain if bacteria possessed genes like those of other organisms,[30] so Lederberg suggested to Ryan that they ought to reverse their approach. "Instead of trying to make *Neurospora* imitate a phenomenon recently worked out in bacteria," they should "inquire whether bacteria had genetic mechanisms similar to *Neurospora*."[31] They would set out to conduct those experiments the following spring.

Chapter 4

Lucky

> ... it occurred to me that no adequate investigation of a genetic nature had been made to demonstrate the existence or absence of sexual recombination in bacteria.
>
> Joshua Lederberg to Edward Tatum, September 19, 1945[1]

Lederberg had the field to himself when he set out to demonstrate genetic recombination in bacteria. Bacteria had been well studied as pathogens since Louis Pasteur and Robert Koch established the germ theory of disease in the 19th century. But bacteria had not been studied from a genetics point of view. To understand why, we need to pause for a moment and understand the place of bacteria in the biological concepts of the time.

Bacteria were thought to be the most primitive organisms on earth.[2] They were not considered to be true cells like those that make up plants and animals, which clearly contain a membrane-bound nucleus with chromosomes. These "true cells" divide in a complex manner called "mitosis": darkly stained chromatin threads grow thicker and straighter to form chromosomes, then the nuclear membrane disappears, the chromosomes move to the center of the cell, and each one splits lengthwise into two chromosomes that move apart toward the poles of the cell where they, in turn, form two nuclei. The cell body then divides by constriction to form two daughter cells.

Bacteria seemed to lack all of this complexity. They had no clearly defined nucleus and seemed to divide through simple fission, not mitosis.[3] It also seemed likely that bacteria lacked sex and did not possess genes like those of other organisms.[4] Botanists who studied them in universities called them the fission fungi, schizomycetes. Pathologists simply called them germs.

In *The Evolution of Genetic Systems*, written in 1939, British geneticist Cyril Darlington referred to "asexual bacteria without gene recombination," and "genes which are still undifferentiated in viruses and bacteria." Julian Huxley summarized what many "knew" about bacteria in his landmark book, *The Modern Synthesis*, in 1942:

> They have no genes in the sense of accurately quantized portions of hereditary substances; and therefore they have no need for accurate division of the genetic system which is accomplished by mitosis… We must, in fact, expect that the processes of variation, and evolution in bacteria are quite different from the corresponding processes in multicellular organisms. But their secret has not yet been unraveled.[5]

This was the conceptual framework when Lederberg set out in the fall of 1945 to see if bacteria exhibited genetic recombination in Francis Ryan's laboratory at Columbia.[6] Their experimental design was simple. First, mix together two mutant strains of bacteria, each of which lacked the ability to synthesize a different amino acid, grow them in enriched media in a flask, and then plate them onto minimal growth media on a petri dish upon which neither could grow, but a genetic recombinant could.[7]

They started some experiments, but their results were inconclusive. (They would later learn that the kind of *E. coli* they had chosen for their experiments would never recombine). Even if they had been successful, their experiment would not have been conclusive because they would have to have addressed an important possible source of error that they had noticed when working on *Neurospora*.[8] Sometimes, a mutant gene could revert the strain to the original normal type strain, and although it might appear to be a case of recombination, it was really nothing of the kind.

Lederberg proposed the term "prototrophs" for strains that have the normal (non-mutant) nutritional requirements. It would be the first term that he added to the scientific lexicon. He tried it out first on George Beadle, the founder of *Neurospora* genetics, in the summer of 1945. "Your term 'prototrophic' certainly has merit," Beadle replied. "Before committing myself finally, however, I'd like to try it out on various of our people."[9]

The appearance of a prototroph due to reverse mutation was something that had to be avoided as a source of error when claiming that recombination occurred after mixing different mutant bacteria strains and then searching for survivors. Lederberg aimed to circumvent the error by mixing strains that each possessed two mutations. Their recombination, their survival, would not be confused with a reverse gene mutation. One gene mutation might spontaneously revert to normal, but it was extremely unlikely that two mutations would do so simultaneously.[10]

The trick was finding such double mutants. Serendipitously, Edward Tatum, Beadle's collaborator at Stanford, had reported that he and his students had isolated double mutants in an *E. coli* strain called K-12.[11] Tatum had acquired the strain from the stock culture collection in Stanford's Medical Microbiology Department. It had been used for student laboratory demonstrations for decades.[12] One of the double mutants required the vitamin biotin and the amino acid methionine for growth; the other required amino acids, threonine and proline. Tatum had his hands on the key to Lederberg's planned experiments to demonstrate recombination.

Then things only got better for Lederberg. The Beadle and Tatum team at Stanford broke up in the summer of 1945 — Beadle moved to the California Institute of Technology in Pasadena while Tatum moved to Yale University. Lederberg wrote to Tatum in September 1945, asking if he had planned to use his double mutants to seek recombination and, if not, whether he would make them available to him. "If an investigation of this sort has already occurred to you, " he wrote, "please let me know, as I am sure that you can do a much better job and have better facilities for it than I; on the other hand, if your plans do not include work such as this I should appreciate very much the service I ask of you."[13]

As it happened, Tatum had no such plans, but he certainly liked the idea and he was willing to send the strains to Lederberg. Ryan had another idea. He knew Tatum, having been a postdoctoral fellow at Stanford learning the *Neurospora* methods the year before. He suggested that instead of sending the double mutant strains to his lab at Columbia, Tatum should take Lederberg to Yale temporarily to work with him.

It was an odd suggestion for Ryan to make. The experiments that Lederberg had in mind certainly could have been done in his laboratory. Ryan would then have been a collaborator and coauthor on some very important papers. Lederberg learned from a colleague decades later that Ryan wanted him to work with Tatum for two reasons: he thought that Lederberg needed someone more senior than himself (a newly appointed assistant professor), and he thought that his being an Irish Catholic from New York was only worse than being a Jew from New York.[14]

At any rate, Tatum wrote to Lederberg on October 15, 1945, inviting him to Yale: "I am hoping that it may be possible for us to work together on some of the problems here at Yale. The questions you raise in your letter have been in my mind for some time, and I have tried to test some of them in a preliminary way with single mutant strains, with rather indefinite results."[15]

The timing was opportune. The Second World War had come to an end the previous month, and civilian life and academic schedules in the United States were soon to be normalized. Lederberg was released from active duty under the U.S. Naval Reserve on Armistice Day, November 11.[16]

Lederberg needed funding, so Tatum suggested that he apply to the Jane Coffin Childs Memorial Fund for Medical Research. Even though Lederberg's research was not directly related to medical application, Tatum thought an exception could be made and that he should stress that his stay at Yale would be important for his career in medical research. Lederberg had a different idea. He saw a potentially important connection between his work on microbial genetics and the biology of cancer.

Cancer cells reproduce in an unregulated manner and Lederberg suggested that this ability might be caused by genetic mutations, which give a body cell a newfound capacity to synthesize an essential metabolite that otherwise would only be available to it in limited and regulated amounts. Tatum thought the idea was "far fetched" and advised the young man accordingly.[17] Lederberg disagreed and published a short note in *Science* titled "A Nutritional Concept of Cancer."[18] It was his first solo publication.

His grant application was successful, but he still needed approval to take time off from medical school at Columbia to do the research at Yale. Then good fortune touched Lederberg again when he received the news that his whole medical class would be given time off to take elective courses for 6 months — from March 20 to September 10, 1946. He immediately wrote to the Dean of the College of Physicians and Surgeons at Columbia requesting that, instead of taking elective courses, he be permitted to work in Tatum's Laboratory at Yale.[19] The Dean's office agreed.

Everything was set and Lederberg was off to Yale on fire. When he arrived in April, there were about half a dozen people working in Tatum's laboratory, including Mary (Polly) Bunting who later became President of Radcliffe and started the Bunting Center.[20] Tatum and the others focused on *Neurospora* genetics, while Lederberg went to work head down on bacteria. Within six weeks, he believed that he had demonstrated genetic recombination in bacteria using Tatum's double mutant strains.

The experiment was elegantly simple, involving two double mutant strains with different nutritional requirements. Strain A would grow on a minimal medium only if the medium was supplemented with the amino acid methionine and the vitamin biotin. Strain B would grow on a minimal medium only if it was supplemented with leucine and thiamine. When the strains were mixed together and if genetic recombination occurred, some of the progeny would possess the right genetic makeup so as to have the ability to grow without added nutrients.[21]

Lederberg's relationship with Tatum was not like his relationship with Ryan. There was no intellectual sparring. Tatum was not a deep thinker, but he was an insightful experimenter with good intuition and great skill in the laboratory, "a green thumb" when it came to growing microbes.[22] Lederberg designed the experiments and did 95% of the work, but he acquired many know-how skills from Tatum.[23]

He got positive results by early June, a week or so after his 21st birthday. He mixed two strains of bacteria, each of which possessed double mutants that were unable to synthesize two specific growth factors. He grew these strains together in a broth, spread them out on minimal growth media on a petri dish, and found evidence of genetic recombination at an extremely low frequency of 1 in every million cells plated. After doing a dozen repetitions that month, he was left with no doubt that the phenomenon was real.[24]

Then another extraordinary opportunity arose. There was to be a major meeting the next month on "Heredity and Variation in Microorganisms" at the famed Cold Spring Harbor Laboratory on Long Island. Tatum was on the program to talk about his *Neurospora* work, and many of the leaders in genetics were going to be there. Lederberg asked if he could attend.[25] There was initially no plan for him to present on his bacterial genetics experiments, but things would change. The young novice would end up giving a spontaneous virtuoso performance defending his work from the fierce criticisms of some of the leading microbiologists of his day.

Cold Spring Harbor 1946

One hundred and thirty American and European biologists gathered for the Carnegie Institution's Cold Spring Harbor Symposium on "Heredity and Variation in Microorganisms" for eleven days, from July 2 to July 12, 1946. The symposia series had become widely known since it began in 1933, but the 1946 Symposium was the first to be held in three years. Due to wartime conditions, researchers had been cut off from research news from other countries.

Much had changed during the War. Genetics had been extended from plants and animals to microbes. Researchers from chemistry and physics crossed into biology and brought new techniques with them. Molecular biology aimed at revealing the nature of the gene itself, its replication, and the steps leading to protein synthesis would soon emerge. That 1946 meeting marked the beginning of a series of Cold Spring Harbor Symposia on genetics, which culminated two decades later in a symposium on the genetic code.

Twenty-five papers were presented in the summer of 1946 with general discussions after each. Talks ranged from the behavior of tumor and leukemia cells to the resistance of bacteria to antibiotics and viruses. For Lederberg, one could not imagine a better setting to meet many of the leaders in genetics.[26] He soon made his presence known, effectively taking over the floor in the discussions that followed presentations.

The same lack of intellectual inhibition and social awareness he had displayed in the classroom at Columbia University was on full display in the meetings with the world's leaders in genetics. He was on fire, speaking fast and furious after every talk. As one participant in the meeting recalled, it was clear he knew his subject well and his mind was damned fast, but if he knew anything about genteel academic protocol, such niceties were not displayed. After dominating the discussions following several papers, the chair of the meetings, recognizing that Lederberg had practically taken over the proceedings, announced that he was nominating him "chairman of the meetings." A bit startled by the comment, Lederberg slowed up for a moment… and then went on as before.[27]

Everyone in attendance at the symposium recognized his brilliance, but many simply found his overwhelming behavior hard to take. Tracy Sonneborn, who pioneered *Paramecium* genetics, took him aside and advised him to tone down his aggressive conduct — to be careful not to unintentionally put others down and to be sure to allow time for others to speak.[28]

The most spectacular moment of the symposium, for many of those who attended, was Lederberg's announcement of genetic recombination in *E. coli*.[29] It happened by accident and rather awkwardly after Tatum's presentation when he made a rather cryptic comment that bacteria would make ideal organisms for biochemical-genetic analysis, except for "their apparent lack" of a sexual phase. One of the listeners was quick to pick up on his use of the phrase "apparent lack."[30] Tatum then referred to Lederberg's experiments. Afterwards, the chair of the symposium arranged for Lederberg to present his data in an informal seminar at the end of the scheduled symposium talks.[31]

It would be the young man's first appearance before the scientific establishment, and for many, it was the highlight of the conference.[32] His claim that he had evidence of recombination in bacteria was intensely and fiercely debated; the discussions lasted several hours. André Lwoff from the *Institut Pasteur* in Paris made the most serious objection. Recall that Lederberg had mixed together two strains of bacteria, each of which possessed double mutants unable to synthesize two specific growth factors. One strain was deficient for growth factors A and B; the other for C and D. He grew these strains together in a broth before spreading them out on minimal growth medium on a petri dish. He found a few colonies that could grow on that minimal growth medium, and he concluded that they were only able to do so because they had genetically recombined and lacked the mutant genes.

But Lwoff offered a counterargument: perhaps they were able grow on minimal medium not because gene recombination had occurred but because of the presence of growth factors in the medium. It was well-known that different bacterial strains could supply each other with growth factor requirements by excreting substances into the medium. If so, the bacteria Lederberg isolated on minimum growth media might have grown because of cross feeding, not cross breeding.

At first glance, Lwoff's objection might appear to be a devastating refutation of Lederberg's claims for genetic recombination in bacteria. But Lederberg gave it no credence whatsoever.[33] He dismissed the possibility that his results were due to cross feeding because a single colony of the bacteria that grew on minimal growth media was stable, and all the colonies isolated from them were also stable.

Lwoff persisted nevertheless, insisting that Lederberg needed to isolate single cells and grow a clone from them, while Lederberg replied that it was superfluous and that he already had other evidence that these were pure clones.[34] The debate became intractable; they were simply talking past one another until microbiologist Max Zelle from the National Institutes of Health intervened and offered to show Lederberg how to isolate single cells to determine conclusively if the ability of the bacteria to grow on minimal media was a genetic phenomenon.[35] The technique was not difficult. Lederberg and collaborators subsequently did it routinely hundreds of times.

Chapter 5

The Road Not Taken

> Ed has suggested that I might be able to get credit towards my PhD here… It might work, but he thought that it was a ticklish business... Could you do the very great favour of sending a thorough opinion of my qualifications for a doctorate in microbiology or genetics to Dean Hartley Simpson…? The sex in bacteria stuff would then be my thesis…
>
> Lederberg to Francis Ryan, August 2, 1946[1]

Lederberg had his work cut out for him when he returned from the Cold Spring Harbor Symposium to Tatum's laboratory in the summer of 1946. He needed more data — more recombinants affecting more traits in bacteria. On October 19, 1946, Lederberg and Tatum sent a preliminary short note to *Nature* titled "Gene Recombination in *Escherichia coli*" to stake their claim.[2] The reaction to their announcement was mixed. Some, like George Beadle, saw it as the most important advance in microbiology since the development of the germ theory of disease. "The story on the sex life of bacteria sounds darned exciting," he wrote to Tatum in September 9, 1946. "It looks to me like it is the most important advance in bacteriology in the last hundred years. Congratulations."[3]

Others were perplexed by the announcement of genetic recombination in bacteria. When yeast geneticist Sol Spiegelman read the note in *Nature*, he thought the claim of recombination might represent some form of word play: "I notice in your letter that you are talking about 'crossing.' Is this a verbal or scientific advance?"[4] Lederberg replied, "I hardly know whether to call 'crossing' a verbal or scientific advance. Hybridization is the only mechanism that can reasonably explain all the facts presented (and some others)."[5] Basking in what he saw to be his new discovery, he sent a copy of his announcement in *Nature* to one of his science teachers at Stuyvesant High School, who replied: "I got quite a kick out of receiving your publication. I got quite a jolt when I tried to read and follow the contents."[6]

Things were looking very good, but as original as his findings were, the evidence for genetic recombination was still relatively meagre. Lederberg had taken only a few months off from his medical school studies at Columbia. Now he would need to ask for more time at Yale while his medical education hung in the balance. In the meantime, Tatum was working on other university business on Lederberg's behalf, negotiating with the administration at Yale to see if Lederberg could be awarded a retroactive PhD.

It was an odd situation. Lederberg was still a medical student at Columbia University's College of Physicians and Surgeons. He was not enrolled at Yale, but he had taken Tatum's seminars in microbiology and biochemistry. He had also attended graduate courses in physiological chemistry, immunology, cytogenetics, and plant physiology. Tatum won approval from the professors of those courses and seminars that Lederberg had de facto enrolled in them. His thesis would be the research on bacterial recombination.

The main difficulty for Lederberg was obtaining funds to retroactively pay tuition to Yale. He thought that there would be a post-war Navy Reserve as part of the V-12 officer program and that he would continue with his service obligations and be paid accordingly. But rather suddenly, and with short notice, the Navy reserve was demobilized in November 1946; the V-12s were simply no longer needed. He had to scramble to find the tuition money elsewhere. But he did receive a New York State War Service scholarship, and with those funds, his savings, and help from his parents, he managed to pay retroactive tuition at Yale.

In the meantime, Lederberg's personal life changed remarkably and for the better. On December 3, 1946, he married Esther Zimmer. She was one of the few women he knew who had a real interest in science. Born in the Bronx, she had received her BA from Hunter College in New York in 1942 before working for two years at the National Institutes of Health in Bethesda, Maryland. She then moved to Stanford where she was a graduate student of Beadle and Tatum working on *Neurospora* genetics.

Lederberg had written to her in the spring of 1946 about her work on a mutant strain in *Neurospora*.[7] She had isolated some of the mutants he got from Tatum to use in his bacterial genetics experiments. But when the Beadle and Tatum team broke up, Zimmer was no longer funded to continue her research with Beadle at Caltech or with Tatum at Yale. Lederberg had met her at the Cold Spring Harbor Symposium where he gave the famous impromptu talk on genetic recombination in *E. coli*. She had just been awarded a Master's degree and was looking for a job. She soon found one at Yale, not far from Tatum's laboratory where Lederberg was working. She was a research assistant for Norman Giles, a cell biologist, who employed her exceptional technical skills to study radiation-induced mutations in *Neurospora*.[8]

Lederberg and Zimmer courted for only three months before he convinced her to marry him.[9] In fact, they were married twice: first in a civil ceremony on December 13, 1946 and again (after informing their parents) in a religious ceremony in which Josh's and Esther's fathers signed the marriage documents. Still, it was far from a traditional Jewish relationship, or far from any other

Fig. 5.1. Esther Lederberg, botanical Garden, Yale University 1948 (Courtesy of the Esther M. Zimmer Lederberg, Courtesy of the Esther M. Zimmer Lederberg memorial Website.)

traditional relationship, for that matter.[10] Their marriage was one in which the home would be run equally, but science would be the primary if not predominant presence in their lives together.

Josh needed more time at Yale. In the summer of 1946, he wrote to the Dean at Columbia College of Physicians and Surgeons asking for another year of leave. He explained the potential importance of bacterial genetics for epidemiology and chemotherapeutic problems in medicine, that at Tatum's request he was registering in graduate school at Yale, and that he would be awarded a doctorate before returning to Columbia to resume his medical studies.[11] His leave was granted.

That extra year at Yale was critical. His aim was to consolidate the preliminary evidence for genetic recombination and start work on a system to map the location of bacterial genes relative to one another using standard genetic methods. To do that, he needed to isolate more nutritional mutants in *E. coli* by subjecting them to X-rays, ultraviolet light, and nitrogen mustard. It was tedious work screening for cells that grew poorly or not at all among thousands and thousands of colonies.

There was still plenty of doubt about Lederberg's evidence for genetic recombination in bacteria. Erstwhile physicist Max Delbrück at the California Institute of Technology in Pasadena was more than skeptical. Delbrück had left Nazi Germany for America in 1937 and a few years later teamed up with Salvador Luria, who had fled fascist Italy for America. They and their collaborators would form what was later called the "phage group," studying bacterial viruses perhaps as "naked genes."[12]

Luria and Delbrück had conducted a landmark experiment in 1943 to see if Darwinian natural selection applied to bacterial mutations. Their data supported the notion that bacterial mutations arose as random hereditary changes just as they did in plants and animals. For Lederberg, those results reinforced the idea that bacteria had genes like all other organisms. But Delbrück and Luria did not see it that way: "Naming such hereditary changes 'mutations' of course does not imply a detailed similarity with any of the classes of mutations that have been analyzed in terms of genes for higher organisms," they said. "The similarity may be merely a formal one."[13] They were equally reluctant to accept Lederberg's explanation that bacteria had sex.

Doubts about Lederberg's work were not just matters of interpreting the evidence. There were also empirical problems in repeating his experiments, at least initially. Luria could not detect recombination when he used a different strain of *E. coli* called "strain B."[14] Their failure encouraged scepticism. But biology is not physics, as different organisms can yield different results. *E. coli* strain B was infertile. Lederberg had been lucky in using

E. coli K-12 for his experiments — only about one strain in twenty would have worked with the protocol he used.

When Lederberg sent out strains of *E. coli* K-12 to Leo Szilard and Aaron Novick at the University of Chicago, he was delighted to learn that they were successful. "I'll eat my hat if this isn't genetic recombination," Szilard wrote to Luria and Delbrück in June 1948.[15] It was not long before others reported that they also had no trouble reproducing Lederberg's results with the K-12 strains he had furnished them with.[16] Strain K-12 would become the model organism for bacterial genetics.

By the spring of 1947, Lederberg had found several other examples of recombination.[17] Luria had finally come around by then to accepting Lederberg's evidence for bacterial recombination and wrote to him with a cartoon of two old men looking in the same direction with the caption, "The trouble with that young man is that he's got nothing whatever in his head but new ideas." "You may hear such type of criticism a number of times," he told Lederberg, "Do not let it upset you: it's just old man's envy — best of Luck to you."[18]

Lederberg passed the comprehensive exam for the PhD at Yale on June 1, 1947 and spent the next two months in the library of the Marine Biological Laboratory at Woods Hole (arguably the best biological library in the world), writing his PhD thesis and exploring the history of microbiology.[19] His thesis was 45 pages long and submitted for publication in *Genetics* on August 1, 1947.[20] It was the first paper on bacteria ever to be published in a genetics journal.

He planned to return to Columbia in September to complete medical school, and Esther would continue her studies as well. In the meantime, they were looking for summer jobs. He saw himself as qualified for genetics, biochemistry, and bacteriology, while Esther qualified for genetics, mycology, and general botany. Jobs were not easy to find. Salvador Luria, who had moved from Caltech to Indiana University in Bloomington, wrote to him in early May 1947 asking if he might know of someone to work with him on bacteriophages the following year. "The salary would be commensurate to training with some mathematics and genetics required, and particularly brains. I wish I could land another Lederberg."[21]

Luria's job was of no interest to Lederberg. Returning to medical school looked to be rather straightforward. Francis Ryan had offered him facilities in his lab to continue working on bacterial genetics for about a third of the time, and the Jane Coffin Childs Memorial Fund for Medical Research was prepared to renew his grant with a stipend of $1,800.[22] But the Dean of the medical school prevented all of this when he wrote to the Child's Memorial fund in mid May, 1947: "May I say that I do not see how Mr Lederberg can devote more than a minimum amount of time to any program outside his third year work in medical school."[23]

At first, Lederberg was devastated, but by the end of summer, the whole issue was resolved — he had made the life-changing decision to drop out of medical school and accept a position as assistant professor of genetics at the University of Wisconsin in Madison.

Chapter 6 Personality Matters

> I believe his ability and brilliance will more than offset his youth (22) and race.
>
> Edward Tatum to R. A. Brink[1]

On April 30, 1947, Tatum received a letter from Alexander Brink, Chair of the Genetics Department at the University of Wisconsin, who was looking to hire someone who worked on the genetics of microorganisms. Tatum had completed his PhD in biochemistry at Wisconsin thirteen years earlier and knew the scene well. "Can you suggest anyone whom we might consider for the post?" Brink elaborated:

> General interest in genetics of microorganisms, is of course, too recent to have attracted many students in this direction. We realize that consequently there is a very narrow field from which to draw… The position is primarily in research, although as the work develops there might be a call for some class room teaching… we would want a man who could work effectively with certain other departments on the campus such as Agricultural Bacteriology, Plant Pathology and Biochemistry.[2]

Brink had knocked on the right door at just the right time. "I have a young man," Tatum replied, "Mr Joshua Lederberg, with me at present who will finish his PhD work this summer. I consider him perhaps the most promising young man yet trained in this field."[3] Tatum explained that Lederberg was planning to go back to finish medical school, but that his interest was in research and whether he returned to medical school would depend on what other opportunities opened for him.[4]

The position in Wisconsin was certainly appealing to Lederberg, but it presented a dilemma. He needed two more years of clinical training and perhaps another two or three of internship and residency to complete his medical education. But that would mean interrupting his research at its most exciting and critical stage, and positions in microbial genetics were rare. Besides, it was far from certain that he would be offered the position.

The genetics department was situated in the agricultural college at the university. It had eight or nine faculty members and was mostly focused on plant genetics, though there had been research on bacteriology in the college itself since the 1920s.[5] Lederberg's background was entirely metropolitan; he had no contact with agriculture at all during his upbringing or during formal schooling.

In fact, most members of the genetics department were not convinced that Lederberg was the right person for the job at all. Most thought that he should not even be invited to interview.[6] But as Chair of the department, Brink was more than interested. Lederberg had sent him the manuscript of an article he had just written describing eight genetic characters, with a sketch of the first genetic map of *E. coli*.[7] Microbial geneticists such as Sol Spiegelman saw it as a masterful work. "I appreciate very much the opportunity to read this," he wrote to Lederberg, "and let me say that I enjoyed every minute of it, not only for the beauty of the experiments, but also for the surprisingly remarkable ability to write."[8] Nonetheless, Lederberg's research was still controversial and there were lingering doubts about it, especially among bacteriologists.

Despite opposition from most of his colleagues in the department, Brink wanted to give Lederberg a chance, so he decided

Fig. 6.1. R.A. Brink, The Joshua Lederberg Papers.

to solicit letters evaluating him anyway. If the letters were favorable, he planned to ask Lederberg to give a seminar and meet the people in the department.[9] The letters came from Yale, Columbia, and the California Institute of Technology, focusing mainly on Lederberg's character. They noted his brilliance and intellectual breadth, but they also warned of his youth, aggressive personality, cockiness, almost haughtiness, and his Jewish ancestry.

Lederberg had been notorious for goading teachers and professors with embarrassing questions they could not answer. He was well known for his extensive knowledge and speed of thinking, but his fierce argumentative style contrasted with the typical polite collegial manner. He almost seemed to go "out of his way to antagonize" his fellow students in Tatum's laboratory when he first arrived at Yale. But Tatum was kind to him, even took him into his home, and as Tatum's other students saw it, "initiated a long overdue process of civilization."[10]

Everyone told Brink a similar story. "Aggressive" was the word used as often as "brilliant". "He is an unusually brilliant investigator, and extraordinarily able," the director of the Jane Coffin Childs Memorial Fund commented. "I had heard that he was aggressive, and saw some evidence of that trait. This seemed natural to me in a man who is so able."[11] Dennis Watson in the Department of Agriculture at Wisconsin advised Brink similarly: "I can fully appreciate the problem with Lederberg," he said. "He has an overbearing air of certainty which makes him many antagonists."[12]

That Lederberg was Jewish was also an issue. As Watson put it, "He is a pretty smart chap, but not the kind to impress people favorably as far as his religion is concerned."[13] Tatum put the matter succinctly when he told Brink that Lederberg's "ability and brilliance" would "more than offset his youth and race."[14]

No question Lederberg had antagonized others in the laboratory, but as Tatum explained, that was "due primarily to his single-minded concentration on science and the problem in hand, which has led to a somewhat thoughtless but only apparently inconsiderate behavior. Even at the worst, however, Lederberg's ability and enthusiasm with his unquestionable personal and scientific sincerity and integrity, have made his colleagues and faculty gladly overlook these lapses."[15] But Lederberg had matured greatly over the past year and "his social sense" had improved considerably. "My frank opinion," Tatum said, "would be that neither this shortcoming nor the only other point against Lederberg, his race, should be taken as serious objection to his consideration for your position."[16]

The Chair of the Department of Botany at Yale, Edmund Sinnott informed Brink similarly. He regarded Lederberg "as the most brilliant graduate student" he had ever met. He not only had a better command of microbiology and microbial genetics than other students, but he also knew "the whole field of biology in a broad way". There was no question for Sinnott that Ledeberg would contribute to the department greatly in terms of ideas. In these regards, he could recommend him "with the warmest enthusiasm." Then he turned to his behavioral difficulties and his being Jewish:

> As you know, he is a Hebrew and during his undergraduate years at Columbia was, I understand, rather insufferable. We have not found him personally objectionable, although when he first came he was inclined to be a little pushful in discussion and sometimes failed to observe all the amenities which most of us value.[17]

But Lederberg had mellowed and grown in every way, Sinnott said, "I would ask for no more courteous or personable fellow than he now is." Actually, Sinnott was less concerned about Lederberg's behavior than he was about the possible behavior of those at Wisconsin towards him: "The fact remains that he is rather obviously Hebraic and in a community which tended to be anti-Semitic he might have a hard time. Tatum knows the Wisconsin scene very well, however, and feels sure that he would be happy there."[18]

As chair of the genetics department, Brink had done his job in soliciting thorough evaluations. As Brink read them, the letters spoke to Lederberg's extraordinary intellect and energy as well as the difficulties he had as a student, "being a little too, perhaps, pushy, forward with his instructors — getting into their hair when he was ahead of them in their thinking."[19] He wrote to Lederberg in July 1947 asking whether he would be willing to visit Madison for interviews with members of the staff and give a seminar on his work.[20]

Lederberg went for the interview completely oblivious to the discussions about his personality and the fact that his being a Jew may have troubled some people.[21] However, his visit seemed to dispel all doubts about him, and he was offered the position a few days later.[22]

A job in an agricultural college was certainly not the perfect fit. He would have preferred a medically oriented environment. But medical application was not going to be a first order of things in his research in bacterial genetics, and there was little interest in genetics in medical schools anywhere in the 1940s anyway. Still, Lederberg asked about the possibilities of cooperating with the medical school at Wisconsin. Brink explained that there was "nothing in particular which would hinder such a development if it seemed desirable."[23] When all was said and done, the position at Madison offered him the best way forward to continue his research.

If there were colleagues at Wisconsin who had some misgivings about taking a Jew onto the campus, they did not betray that to him and he was treated completely cordially.[24] But no one welcomed him more warmly than Brink. It was an avuncular relationship, Lederberg recalled decades later. "He was someone old enough, almost, to be my uncle, and in some respects treated me as such... But, anyhow, these folks just dealt with me so kindly, so generously."[25]

There were all kinds of impediments that all faculty members had to face in the fall of 1947. They were working in difficult circumstances just after the end of the Second World War. There was an overwhelming flood of students, and they did not have the space, equipment, or staff to take care of them properly. As Brink recalled, "Josh realized what the situation was and he was determined to take his share of the responsibility in working it out. I always had great admiration for him."[26]

Lederberg caused no problems at Madison. "All the dire things that people predicted would happen when he came to our campus didn't happen," Brinks later said.[27] In fact, he was a catalyst for change and fresh ideas just as his professors at Yale said he would be. His undergraduate course was attended by many faculty members as well as students.[28] He helped shape not only the genetics department, but also the development of biology as a whole at the university. Brink considered his role in bringing Lederberg to Madison as one of the most important achievements of his career.[29]

Lederberg did not learn about the discussions of his being Jewish until many decades later when he asked Brink to send him the correspondences pertaining to his appointment. He was startled at the discussions about "race" and the confusion between the individual and the group. The letters were intended to overcome prejudice, but they still attributed his personality to his racial background.[30] What people had regarded as his aggressiveness was due not to his "race" but to his perception of the logic and rigor of science. "When somebody was speaking nonsense, I would show it up, and that's a fault… You didn't dare make a mistake in my presence, because I would pounce on you, OK? Well, that could be a fault. I can sympathize with people being put off by that."[31]

Lederberg's cockiness and aggressiveness continued to be an issue when Warren Weaver of the Rockefeller Foundation visited him in the winter of 1948. The Rockefeller Foundation played a critical role in funding the development of microbial genetics and in "molecular biology," an expression coined by Weaver himself in 1938.[32] Brink had informed Weaver of Lederberg's arrival at Wisconsin several months earlier. "Lederberg is an unusually able young investigator, with an extraordinary breadth of knowledge and biological interests," he wrote to Weaver. "He is beginning to be recognized already as a spark plug in the biological research group on the campus. We are hopeful of important things from him."[33]

Weaver was impressed with Lederberg's intellect, but not so much with his unkempt appearance: "He is a little fat and somewhat sleepy and sloppy looking," Weaver wrote in his diary. "But the minute he starts to talk, you forget all this. He speaks very well indeed — clearly, modestly, but very forcefully."[34] He was awarded a grant of $7,500 for three years. The "exceedingly brash and aggressive young man" had vanished when Rockefeller officials visited Lederberg a year and a half later. He had "matured rapidly" and was turning out "first rate work."[35] He was just 24 years old.

Lederberg remained at Wisconsin for twelve years. Together with his wife, Esther, and a small group of students, he established the methods and concepts that came to define the new field. In those years, bacterial genetics moved from the realm of doubtful science to the center of biology and allowed for the birth and explosive growth of molecular biology.

Chapter 7

A Field of Their Own

> I just wanted to get to my work — I had a lot to do in the lab. It was, in fact, a very fruitful, very busy time.
>
> Joshua Lederberg on laboratory life in Wisconsin[1]

The genetics building in the College of Agriculture looked like an over-grown log cabin, and when Lederberg arrived at Madison during the fall of 1947, his laboratory was a small 20 × 30-foot room in the basement.[2] The university remodeled a new laboratory of about the same size for him by the spring of 1949. It was on the second floor right under the eaves and crowded with glassware, autoclaving and media preparation equipment, and a few benches. Laboratory space was just not that important for experimental breeding, which was the main focus of the genetics department then. As most of that work was in the field, their laboratory was outdoors.

Although Lederberg's laboratory was small, he essentially had the whole field of bacterial genetics to himself, and he formed a small but talented and industrious group with his wife, Esther, and a few graduate students. In rapid succession over the next few years, they announced one breakthrough after another that included discoveries of hitherto unknown hereditary phenomena as well as new methods that effectively transformed bacterial genetics into one of the most vibrant and important fields of biology. Needless to say, Josh orchestrated the work, determined what experiments were to be done and how to do them, and he devised the methods, and was also heavily involved in interpretations of the results. As Lederberg's first and most famous graduate student, Norton Zinder, put it, "You can credit most ideas in the lab to Josh."[3]

Recently, who did what has become a subject of controversy. Much of it concerns the role of Esther Lederberg. Since her death in 2006, there have been suggestions that she deserves much more recognition for the research in the laboratory than she received — that she had been overshadowed by her husband's fame.[4]

When the Lederbergs arrived at Wisconsin, Esther immediately began to pursue a PhD. The chair of the genetics department,

Fig. 7.1. Joshua Lederberg in the laboratory, The University of Wisconsin, 1958, The Joshua Lederberg Papers.

Fig. 7.2. Esther and Joshua Lederberg, 1947. (Courtesy of the Esther M. Zimmer Lederberg memorial Website.)

Alexander Brink, was ostensibly her thesis director. But as he was a corn geneticist, her actual work was on bacterial genetics under her husband's supervision. He suggested a topic closely related to Brink's interests on the genetic control of mutability. She aimed to determine whether there were genes affecting mutability in *E. coli*. She completed her doctoral degree in 1950.

The state of Wisconsin had strict nepotism rules instituted during the Great Depression that prevented two family members from working as staff members in the university. But the nepotism rule was relaxed somewhat by the time Esther finished her doctoral degree; she was able to get a dispensation. Brink found funds for her by placing her name on an army contract involving state farmers.[5] She worked in her husband's laboratory as a "Research Associate" for the next eight years.

Replica Plating

Commentators on Esther's career have recently suggested that not only is she underappreciated as a great scientist, but that her husband was given credit for work that was rightfully hers. So it is claimed, for example, that it was she, not her husband, who invented "replica plating," one of the most innovative and important methods in microbial genetics.[6] However, that claim may be fundamentally misguided.

Replica plating is a method for handling many clones simultaneously. It was designed to replace the tedious job of hand picking and testing one bacterial clone at a time in genetic work on. The usual procedure for testing a bacterial colony was to use an inoculation loop: a thin platinum rod with a looped wire at the end that can hold a drop of liquid — somewhat like a bubble wand holds liquid soap. A bacterial colony was "scooped up" and put onto a new agar plate to see how it grows on a medium lacking nutrients or one containing growth inhibitors, such as an antibiotic.

Transferring colonies from one kind of media to another was a chore, and one had to sterilize the device repeatedly. Lederberg used a beaker full of sterile toothpicks instead of having to flame the platinum loop between picks. Still, it was tedious. Aaron Novick and Leo Szilard at Chicago had invented a little device with steel prongs on a wire brush so that one operation would serve as multiple transfers of a single needle. That way, you could get a hundred colonies transferred all at one time. But that method was still cumbersome and did not work well because the points had to be lined up exactly right. Some used paper, but it smeared the colonies.

Josh Lederberg had another insightful idea to use a fabric with a vertical pile, which would be an analog of the wire brush and the paper. He read books on fabric structure and collected a wide variety of remnants from a dry goods store to test them.[7] Cotton velveteen worked perfectly because the pile provided space for moisture that might otherwise cause smearing of any impression. So, one could take an impression of the colonies from one master dish and reproduce the spatial pattern of the colonies on another dish perfectly.

"Replica plating" was of enormous technical help for many bacteriological experiments.[8] The Lederbergs first used it to test the origin of bacterial mutants, a vexing issue in bacterial genetics. When bacteria colonies are grown in the presence of an antibiotic (or even virulent viruses), some cells survive. The question was, were those antibiotic-resistant strains induced by the presence of the antibiotic or were they in the population before the antibiotic was added? Replica plating afforded a direct test: First, grow a plateful of bacteria from a small inoculum of a few thousand cells and let them multiply to a billion. Next, make a plate with many distinct colonies on it. Then use the velveteen to make an identical plate, add the antibiotic to each plate, and search for survivors. If the surviving clone appeared in the same geographical location on each of the plates, then one could conclude that mutation was there prior to adding streptomycin.

Josh was by no means always great in the lab. He was sometimes clumsy, especially with glassware. But he was famous for the rate at which he generated ideas, his encyclopedic knowledge, and his ingenious and simple experimental designs for solving profound questions. His was a barebones approach. As one collaborator quipped, "Josh's rule was that if an experiment required more than 6 Petri dishes and four pipettes, he felt it was over designed."[9] Replica plating was exemplary. As one of his colleagues wrote to him in 1951, it had "a classic simplicity; and the results obtained with it seem perfectly clear-cut."[10]

Decades later, however, rumors spread that using velveteen was actually Esther's idea and that it was she who had picked out the fabric. The method for replica plating had Esther "all over it," it was said. "After all, she was a practical scientist who came from a family of textile workers."[11] It is not clear what "practical scientist" means, but Josh had long before debunked such stories as possibly tainted with sexism. Esther's expertise was not in clothing materials, he said, but in actual performance in the laboratory. She was a "superb person in the experimental manipulation," he said, "a very meticulous experimentalist."[12]

Lambda

Nothing exemplified Esther Lederberg's keen and careful observations and laboratory skills more than her research that led to the discovery that a virus could live symbiotically within its host. During the second year of her doctoral thesis research, Esther astutely noticed holes that appeared on bacterial lawns on plates of *E. coli* strain K-12. She had no idea where they came from and thought it was contamination at first, but Josh suggested that she follow that up. As it turned out, the standard strains of *E. coli* were carrying a bacterial virus (phage) that had been maintained in stocks of K-12 since 1944, but had remained undetected. Esther reported on the existence of the phage in January 1950.[13] Then, using the cross-breeding techniques that Josh had devised, she studied the phage as a genetic factor.

The Lederbergs named the phage *Lambda*, because they had initially thought it was a symbiont, like the genetic entity called

Kappa that reproduced itself in the cytoplasm of *Paramecium*. *Kappa* was normally transmitted sexually and from one cell generation to the next in *Paramecium*, but it was also shown to be infectious through laboratory techniques. It was determined to be a symbiotic bacterium and became a model for understanding hereditary symbiosis. But in 1953, the Lederbergs reported that the analogy with *Kappa* was not quite correct, because Esther had shown that under certain conditions, *Lambda* DNA could become incorporated into the main bacterial genome.[14]

The idea of hidden viruses coexisting in symbiotic association with their bacterial host had been poorly understood and not well accepted for decades.[15] During the 1920s and 1930s, Eugene and Elizabeth Wollman, working at the Pasteur Institute in Paris, reported that phages could actually enter into a latent form in their host cells, and in such a way that they might be passed harmlessly from one cell generation to the next. When the bacteria are stressed, their cell membrane breaks down and the bacteriophages are released. Those cryptic bacteriophages were referred to as lysogenic. Tragically, the Wollmans were seized by the Nazis in 1943, sent to Auschwitz, and never heard from again. But in 1949, André Lwoff confirmed the Wollmans' work and in 1952 gave the name "prophages" to lysogenic phages that are integrated into their host genome. *Lambda* in *E. coli* K-12 was one such prophage.

Lambda not only became the model system for studying viruses and virus-host interactions, but was also intensely studied in many laboratories to understand fundamental problems in molecular biology: how DNA is replicated, how genetic information is transcribed to RNA and then translated into the amino acid sequence of a protein, how genes are turned "on" and "off."[16]

New Methods for Mutants

In the meantime, another remarkable phenomenon was discovered in the Wisconsin laboratory in the early 1950s: viruses could actually pick up host genes and transfer them to other bacteria they infect. Startling evidence for that surfaced when Lederberg extended his research program from *E. coli* K-12 to Salmonella, whose genetics he had been interested in before he worked on *E. coli* K-12 at Yale.

Salmonella had been studied in detail because of its public health implications in food poisoning, typhoid fever, and other intestinal infections. Pathologists had identified types of Salmonella based on their particular antigens, the proteins that host antibodies recognize and respond to. Antigens provided a way of tracking strains through epidemics. Every antigenic strain of Salmonella was given a binomial name to commemorate a place (e.g., "*Salmonella newport*"). Taxonomic tables of diverse Salmonella strains were published, and when Lederberg read those tables, he could see different combinations of antigens which hinted that genetic recombination was somehow occurring.[17] He turned to Salmonella genetics as soon as he had his own laboratory.

Lederberg once said that discovering transduction was like finding "a big fat worm under a rock", and that there were three aspects to finding that worm. The first involved training Norton Zinder, a young novice. The second was inventing a new method for obtaining mutants. The third was in coming to grips with the idea that a virus was involved in its host's heredity.

Zinder was an unruly albeit very bright 19 year-old when Lederberg took him into his laboratory. Though they were only three years apart in age, their relationship was more like father and son than brothers. Lederberg nurtured and trained Zinder and set him out on a path that led him to become one of the most outstanding microbial and molecular biologists of his generation.

Francis Ryan recommended Zinder to Lederberg in June 1948. At that time, he had just completed his bachelor's degree in Zoology at Columbia. Ryan considered him to be "a first-rate man" who "stood out academically at the head of his group." He had been applying for medical school since he was 18 years old, but had been repeatedly turned down because he was too young and "a bit too sure of himself," Ryan said, and he was planning on reapplying to medical school as soon as he could.[18] Ryan found him to be "very personable" and thought he had "the ability,

Fig. 7.3. Norton Zinder 1953. (Courtesy of the Esther M. Zimmer Lederberg memorial Website.)

enthusiasm and training to enable him to develop into a top-notch investigator."[19]

Zinder arrived at Madison in July 1948. Lederberg's plan was to eventually assign him the task of extending the genetic methods he had devised for *E. coli* to Salmonella. But first he needed training, so Lederberg put him through the ropes, teaching him how to handle bacteria and especially a pathogen as dangerous as Salmonella. He spent the summer making media, cleaning glassware, and learning to isolate mutants of *E. coli* K-12 for genetic analysis.

That last learning step of isolating mutants in *E. coli* K-12, as it turned out, was critical for studies of Salmonella. Isolating mutants was like searching for a needle in a haystack.[20] After the first six weeks or so, Zinder could not find a single one. He was frustrated and felt defeated. Then in August, Lederberg came up with an ingenious way of finding them. He remembered learning in medical school that penicillin only works on bacteria that are growing. He went to the library to find the paper.[21]

At first glance, one might not see how this knowledge about penicillin could apply to finding mutations, but thinking laterally was one of Lederberg's great strengths. If one placed a culture of bacteria that was treated with a mutagen, say ultraviolet irradiation, on minimal growth media that also contained penicillin, mutants that could not synthesize a particular nutrient would not grow and, thus, they would be the only ones to survive. Those mutated bacteria could then be plated in media that contained the necessary nutrients and growth would resume. The method worked!

Within a few days, Zinder had isolated more biochemically deficient mutants than he could analyze in a lifetime.[22] He and Lederberg published a paper on the new methods in 1948.[23] Like replica plating, it would become a classic paper and once more exemplified Lederberg's talent at finding simple solutions to address difficult problems. As one admirer wrote to Lederberg about their penicillin treatment paper, "Like so many classics, once the solution is pointed out, it seems simple... The trick of course is to point out the solution."[24] The use of penicillin to identify and isolate biochemically deficient mutants in bacteria was a great technical achievement that accelerated bacterial genetic research.

Zinder would apply those methods to Salmonella. At first, he still had his sights on medical school, but six months after arriving in Wisconsin, he was turned down again by Columbia University. So, he remained in the lab as a graduate student and began work on Salmonella. Stocks arrived in a diplomatic pouch from Stockholm in early February 1950.[25]

Using the penicillin enrichment technique combined with UV mutagenesis, Zinder isolated many mutant strains and tested them for genetic recombination by mixing two cultures, plating on a minimal medium, and looking for colonies that could grow, just as Lederberg had done for *E. coli* K-12.[26]

Lederberg was very strict about obtaining complementary (double) mutants of Salmonella to use on both sides of the cross, just as he had done when demonstrating recombination in *E. coli* (Chapter 4). One mutant would lack the ability to grow without factor A in the medium; another could grow without A, but not without factor B. The reason he wanted double mutant strains, it may be recalled, was to ensure that any bacteria that were found capable of growing on minimal media, after mixing them, were due to genetic recombination and not spontaneous gene reversion. If only a single nutrient deficient mutant is used, it could revert spontaneously and grow on minimal media, and Lederberg wanted to avoid that source of error.

The Contraceptive Experiment

Finally, in the fall of 1950, Zinder found a particular double mutant that could be mixed with another double mutant. One strain had mutations that caused it to require the amino acids methionine and histidine; the other strain required the amino acids phenylalanine and tyrosine. He observed growth on minimal media lacking those four amino acids. It was clear that genetic recombination occurred which allowed those mutant cells to survive, just like *E. coli* K-12.

But there was a complication. It was not clear whether cell-to-cell contact was involved like it was with genetic recombination in *E. coli* K-12. When strains of *E. coli* K-12 colonies were grown in a U-shaped tube separated by a filter that allowed fluid but not bacteria to go through, recombination did not occur.[27] This indicated that cell contact was required, as Lederberg concluded when he referred to it as conjugation. However, recombination did occur when Salmonella strains were tested in the same way. Something was getting through the bacteria-proof filters that led to genetic recombinants. They called the unknown filterable agent "FA" for short.

The genetic effect of FA was something like the transforming principle in pneumococcus, but Zinder and Lederberg ruled out naked DNA because the enzyme DNAse, which degrades naked DNA, did not alter the activity of FA. It was possible that FA could be a special "filterable form" of bacteria. Lederberg knew of reports published since the 1930s that certain kinds of bacteria, when subjected to noxious conditions, could morph into protective granular forms that were capable of passing through filters. They were called "L-forms" after the Lister Institute in London where they were discovered.[28]

The idea that FA could be a virus, which picked up bits of host DNA and carried it to other hosts, did not occur to them until they attended the Cold Spring Harbor Symposium in the spring of 1951. It was a tumultuous meeting. Genetics was in a revolutionary state as old paradigms were coming undone and new ideas were rapidly hatching. It was still not yet certain if genes were proteins or DNA. Proteins had been the favorite candidate because of the assertion by chemists that DNA was too simple. But that was starting to be questioned.

The one gene-one enzyme hypothesis of Beadle and Tatum was also coming undone, with evidence that gene-enzyme relations were far more complex: several enzymes may be affected by a single gene and, inversely, several genes could affect a single enzyme. The world's leading cytogeneticist, Barbara McClintock from the Cold Spring Harbor Laboratory, gave a talk arguing that genes were not static entities as some could move around and regulate gene expression. It was uncertain how many people really understood her talk.[29]

But as Zinder remembered it, Lederberg's talk "won all competition for incomprehensibility."[30] It also won the competition for length. He spoke for a record-breaking six hours, dealing with four years' worth of data from his Wisconsin laboratory: the genetics of *E. coli* K-12 and the unknown mechanism underlying it, the relations between genes and enzymes, *Lambda* phage, and the mysterious results on genetic recombination in Salmonella.[31] British geneticist Charlotte Auerbach remembered him at the meeting as "a very young, very bright, very arrogant, and very likeable chap, who talked from 7PM to 1 AM."[32]

Bacterial genetics was indeed in a confused state. However, as Zinder related the story 40 years later, he got a lead on the nature of the filterable agent in Salmonella when he spoke with Harriet Ephrussi-Taylor at that Cold Spring Harbor meeting.[33] She suggested to Zinder that it could be "a virus with some DNA stuck onto it."[34] Led astray by L-form bacteria, he and Lederberg had not considered that possibility. That also meant a clear line of new experiments when they returned to Wisconsin. Everything pointed to the idea that FA was indeed a phage.[35]

Zinder and Lederberg published their first joint paper titled "Genetic Exchange in Salmonella" in the spring of 1952, introducing the term "transduction," which means to lead across. The virus, they said, "would function as a passive carrier of the genetic material transduced from one bacterium to another."[36]

Zinder was in demand when he completed his PhD in the spring of 1952. Rollin Hotchkiss approached him about an opening in his laboratory at The Rockefeller Institute for Medical Research in New York. Hotchkiss worked on "the transformation principle" in pneumococcus which, in his view, was indeed DNA.[37] The chemists' claim that it was a boring, monotonous molecule was erroneous. He had reported chemical data indicating that DNA varied between species and that purified bacterial DNA could transfer penicillin resistance from one strain of pneumococcus to another.[38] He wanted Zinder to work with him on viral transduction in Salmonella involving "protected DNA." Zinder joined the Rockefeller that year and remained there for the rest of his distinguished career.

After Zinder left Madison, Lederberg turned to the genetics of Salmonella antigens, which had first piqued his interest in bacterial recombination 7 years earlier.[39] In the mid-1950s, Larry Morse together with the Lederbergs reported that transduction also occurred in *E. coli* K-12, and that *Lambda* phage could also pick up host genes and transfer them to new hosts. But unlike transduction in Salmonella, *Lambda* transferred only a few host genes, specifically those near to the site where it is incorporated in the host's genome.[40] They called it "specialized transduction."

The evidence that viruses could pick up host genes and transfer them to other bacteria presented a whole new mode of heredity that was virtually unheard of before these reports. It was fundamental for understanding not only bacterial genetics, but also the origins of genes themselves. Perhaps many genes, including some of our own, originated as infectious viruses. Joshua Lederberg theorized on this and on what he called "infective heredity" in a landmark paper in 1952 (Chapter 9).

But transduction was not the only form of this newly discovered "infective heredity" in bacteria. As it turned out, an infectious fertility factor also permitted bacterial conjugation to occur in *E. coli* K-12. The struggle to reach a consensus on the nature of bacterial sex, if it could be called that, was one of the most complex and difficult challenges in the emerging field of bacterial genetics.

Chapter 8 Sex Controversy

> Despite Joshua's fabulous cranium, the genetics of bacteria became messier each year. Only Joshua took any enjoyment from the rabbinical complexity shrouding his recent papers. Occasionally I would try to plow through one, but inevitably I'd get stuck and put it aside for another day.
>
> James Watson, 1968[1]

Understanding bacterial sex was not easy as it differs starkly from that in plants and animals. There are no gametes that fuse to produce fertilized eggs. During bacterial conjugation, genes are transferred in a unidirectional fashion from donor cell to recipient cell. But all bacterial cells are not able to transmit genes this way. Only those that carry a "fertility factor" (F^+) have this ability. F^+ bacteria can transfer genes to another bacterium that lacks the fertility factor, but even then, they only transmit some of their genes, not all of them.

The fertility factor itself is a genetically autonomous entity that is usually located in the cell cytoplasm of bacteria, and it can be transferred to F^- bacteria by infection, turning them into gene donors. Only rarely is F located in the main bacterial "chromosome" or genophore. There was nothing comparable to this system in any other kind of organism.

Establishing this consensual view of bacterial sex emerged from a controversy between the Lederbergs and Luca Cavalli on the one hand and William Hayes on the other. Neither of the rivals got everything right. There were two controversial issues. One was over the fertility factor itself — was it an autonomous entity that conferred compatibility and permitted gene transfer from donor to recipient, or did it actually carry the genes that are transferred to recipient cells like, say, how a sperm cell does? Lederberg and collaborators held the first (correct) view, Hayes the second.

The other vexing issue was why only some genes from donor cells are found in the progeny of the fertilized cell. Everyone agreed that the genetic contribution of the F+ cell to progeny was typically much less than its complete genome. How did this happen? Did a reduction in donor gene numbers occur after fusion with the F- cell, or were only some of the donor cell's genes transferred to the recipient cell in the first place? Lederberg and collaborators held the first view while Hayes held the second (correct) interpretation.

To understand how the debates originated and how these issues were resolved, we need to back up a little in our story. When Lederberg wrote an overview of his work in 1947, he concluded that his results "provide evidence of an organized array of genes comparable to that of higher forms."[2] His evidence for bacterial conjugation seemed to be strong, but neither he nor anyone else knew anything about the actual physical process by which conjugation occurred. He suspected that cells of different sex types fuse and form a zygote, a fertilized egg of some kind, like in other organisms. But it was all guesswork. Max Delbrück wrote to him fiercely on February 26, 1948 saying that unless the actual physical process by which genes are exchanged could be shown, he rejected all of his "bacterial genetics":

> I refuse to believe in your bacterial genetics (and therefore also refuse to take an interest in details of the type you mention) as long as the

kinetics of the postulated mating reaction has not been worked out. After our discussion last summer I thought you were going to do the obvious experiments to clear this up.[3]

Lederberg had no idea what experiments Delbrück was referring to, nor did Delbrück himself as it turned out. Mating or recombination was a rare, one in a million event and could only be detected in large populations.[4] Actually observing cells joining up with one another under the microscope, if that is what they did, was like finding one needle in a haystack of a million identical needles.[5]

After Esther Lederberg completed her thesis in 1950, she and her husband worked together on understanding conjugation in *E. coli* K-12. At first they ran a blank, but then a number of things started happening in very quick succession. One was the discovery that there were "mating types" in *E. coli* F^+ and F^- cells, though this was not a great surprise because there were already mating types designated as plus and minus in the bread mold *Neurospora crassa* as well as other microbes as revealed by subsequent studies. Crosses occur only when + and – strains are mixed together. There is no obvious morphological difference to indicate which is which. One could not be spoken of as being "male" and the other as "female"; they were arbitrary alternative types.

The Lederbergs teamed up with Luca Cavalli, who provided the first independent verification of genetic conjugation. When the Lederbergs first sent him strains of *E. coli* K-12 in January 1949, he was working at the University of Cambridge in the laboratory of well-known population geneticist and mathematician Ronald Fisher. Fisher had read Lederberg's 1947 paper providing evidence of the first genetic map of *E. coli* and wanted to develop the new field.[6] He met Cavalli a year later at a conference in Stockholm when Cavalli gave a talk on bacterial resistance to radiation and other potential mutagens.[7] At that time, Cavalli worked at a pharmaceutical firm in Milan, the *Instituto Sieroterapcio Milanese,* one of the few research institutions open in Italy after World War II. Following his talk in Stockholm, Fisher offered him a job to pursue bacterial genetics in his laboratory, which had mainly focused on mouse genetics.[8]

Lederberg was only delighted when Cavalli wrote to him from Cambridge in the fall of 1948 about his plans to set up a bacterial genetics lab. "I will of course assist you in any way that I can, and correspond with you on any detail," he replied.[9]

Cavalli easily produced what was one of the most significant confirmations of Lederberg's experimental results on genetic recombination in bacterial K-12. But as Lederberg recalled years later, "the skepticism around me was incredible. To classical bacteriologists, we (the very few bacterial geneticists) were lunatics."[10] In fact, Fisher failed to find funds to support his new genetics program and it was eliminated after two years. Cavalli returned to his old job in Milan where he was permitted to work part-time on his genetic research.[11]

Cavalli and the Lederbergs developed a close and friendly relationship exchanging stocks of mutants, data, and ideas. They were in sync. What one laboratory was planning to do, the other was about to do or had just done. But a competitive context arose when William Hayes turned to work on mating in *E. coli* K-12. Hayes had graduated from Trinity College, Dublin and then obtained a medical degree in 1937. He later developed an interest in bacteriology when stationed in India during World War II where he worked on Salmonella infections and penicillin therapy.

Hayes was working at the University of London Medical School when Cavalli met him in a summer course on bacterial chemistry in Cambridge in 1951. Cavalli gave him strains of *E. coli* K-12 and showed him how to make genetic crosses. Hayes then focused on the mating process underlying recombination and entered into correspondence with both Lederberg and Cavalli on the nature of the fertility factor F and the process underlying conjugation.

There is considerable confusion today about this history. It is sometimes stated that Hayes was the first to discover the fertility factor F.[12] But such statements are erroneous. Indeed, as we shall see, Hayes actually opposed the idea of independent fertility factor F that permitted the transfer of genes from donor to recipient and instead proposed that the fertility factor itself carried the body of genes that were transferred from F+ to F– cells.[13]

Thinking he had beaten Lederberg at his own game, in November 1951, just a few months after he met Cavalli, Hayes quickly sent a letter to *Nature* with the title, "Recombination in Bact. Coli K-12; Uni directional transfer of genetic material."[14] It appeared in the January 19th issue. Therein, he suggested that the male cell extruded a "gamete" that was taken up by female cells on contact. He suspected that the male gamete was a virus, perhaps *Lambda*.[15] In December 1951, he sent Josh a draft of his letter to *Nature*, while requesting that Lederberg send him a number of mutant cultures of K-12.[16]

Hayes's letter and his involvement in bacterial genetics came as an unhappy surprise to Lederberg. He thought that he had the field to himself, and thus the luxury of time to work out the nature of genetic recombination in an unhurried, contemplative way. But Hayes's letter to *Nature* changed all that. Lederberg sent him stocks and entered into cordial discussion while at the same time trying to dissuade him of his faulty interpretations, first that the fertility factor "F" was the phage *Lambda* which transferred genes, and second, that it was an infectious "microgamete" analogous to sperm that carry genes and infect female cells.

Lederberg long had evidence that genetic recombination in *E. coli* K-12 required cell-to-cell contact and was not a matter of viral transduction as in Salmonella, for example. The Lederbergs had considered the idea that *Lambda* might play a role in fertility but ruled that out because there was no correlation between genetic recombination between strains and the possession of the *Lambda* phage. Strains that lacked *Lambda* were capable of being genetically crossed.[17] Josh wrote to Hayes in January 1952 explaining the situation.[18]

That month, Josh and Esther also conducted experiments that indicated that the fertility factor is an infective agent that could be transmitted through the medium and transform F^- to F^+ cells. Cavalli did the same. Hayes was going to conduct the same experiment and was flabbergasted when Cavalli informed him on February 10, 1952 that he and the Lederbergs had done it a month earlier.[19] Insisting on his own independent discovery and priority, he replied to Cavalli, "I was most interested to receive your letter of 16 February and to learn that both yourself and Lederberg had arrived at the conclusion that an infective agent was concerned. Your predictions (and my own!) as to the results of my experiments on infection are correct."[20] Cavalli and the Lederbergs published their results on "sex compatibility" and the infectious nature of the fertility factor in 1952.[21]

But there still remained a fundamental difference between them and Hayes. Hayes disagreed that fertility factor F was a distinct genetic entity that conferred sexual compatibility and permitted genetic transfer. He insisted that the fertility factor was actually an infectious "microgamete" that carried genes from donor to recipient. As Lederberg saw it, Hayes was muddying the water. He wrote to Hayes on February 27, 1952, cautioning him about confusing fertility factor F with *Lambda* and/or conceiving of it as a "self-reproducing gamete" that carries genes from donor to recipient. But Lederberg was only partially successful. Although Hayes did eventually drop the idea that *Lambda* was involved, he continued to insist on an infectious virus-like "gamete" that carried genes from donor to recipient.[22]

The Lederbergs and Cavalli agreed to publish their views and those of Hayes back to back in 1953 in the *Journal of General Microbiology*.[23] Their differences in interpretation remained sharp. Hayes maintained that "The F+ agent thus acts as a gene carrier in the transfer of genetic material from F+ to F– cells."[24] The Lederbergs and Cavalli maintained that F was a compatibility factor and not "a gene carrier in the transfer of genetic material."[25]

By that time, Hayes' interpretation of unidirectional gene transfer by an infectious gamete had attracted much attention, especially after he gave a presentation at a symposium on microbial genetics in Pallanza on the shores of Lake Maggiore, Italy, in September 1952. Many leading microbiologists were impressed with his idea of a microgamete that carried genes unidirectionally to recipient cells.

Hayes's presentation also got the attention of the brash and bright 24-year-old James Watson, another wunderkind who would soon become famous for his discovery with Francis Crick of the double helical structure of DNA. Recalling this period years later in his blockbuster book and personal account of the discovery of DNA structure, *The Double Helix* (1968), Watson wrote of

Hayes's presentation in Italy: "Bill's appearance was the sleeper of the three day gathering... As soon as he finished his unassuming report, however, everyone in the audience knew that a bombshell had exploded in the world of Joshua Lederberg!"[26]

After hearing Hayes's presentation, Watson decided to team up with him to further best Lederberg. Watson had just completed his PhD the year before and was spending a lot of time in Cambridge with Crick working on the structure of DNA. He visited Hayes when he went to London for discussions on the X-ray diffraction analysis of DNA, which Rosalind Franklin had famously carried out.[27]

The Double Helix was a remarkably candid but flawed account of the race to discover the structure of DNA, and it threatened to end Watson's friendship with Crick, who considered it "a stab in the back," "an invasion of his privacy," and "in bad taste."[28] Lederberg thought the book displayed some of Watson's "egregious idiosyncrasies," and that Watson's self-caricature of ruthless competition conflicted with his own view of science as a "higher calling," a search for truth that transcended "personal, national, ethnic, and sectarian rivalries."[29]

Lederberg and Watson were indeed very different personalities. Hayes later recorded his impression of both men in his autobiography: "My first impressions of Jim suggested a naïve gaucherie tinged with arrogance… Jim's lanky figure occupied three seats of the front row of the theatre, his legs resting on two of them… I later got to know Jim well and found him highly intelligent and enjoyed his company, despite, or perhaps because of his lack of maturity."[30] Hayes's impression of the 28-year-old Lederberg differed dramatically when he first met him in Madison in the summer of 1953 *en route* to Caltech to work for a few months in

Delbrück's laboratory[31]: "Joshua displayed a portly figure matched by a powerful personality and an aura of deep intelligence. I was told that his bedtime reading was a treatise on advanced mathematics by Sir Arthur Eddington."[32]

The Double Helix also included an account of Watson's attempt to top Lederberg, who, he said, had the field to himself "as the *enfant terrible*" and had carried out such "a prodigious number of pretty experiments that virtually no one except Cavalli dared to work in the same field."[33] "Particularly pleasing," he wrote, "was the possibility… that I would accomplish the unbelievable feat of beating him to the correct interpretation of his own experiments."[34] He and Hayes began a collaboration to do just that.

However, they only confused things further when they offered a new interpretation of Lederberg's genetic mapping data which, Lederberg said, indicated nonlinear interactions among three sets of gene clusters. They incorrectly proposed that the three linkage groups were evidence of three distinct chromosomes.[35] As it turned out, the main genome of *E. coli* consists of a single circular "chromosome."

Hayes's denial of an autonomous fertility factor and his talk of a "self-reproducing gamete" aside, what remained in doubt for Lederberg was why the genetic contribution of the F^+ cells to the progeny was less than that of the F^- cells. Was it because the donor cell transmitted fewer genes prior to zygote formation as Hayes believed, or did the reduction of donor genes occur after zygote formation, as Lederberg proposed? Lederberg had long considered the former hypothesis himself, but dismissed it as it was difficult to reconcile with some of the (albeit hardly decisive) evidence. Still, Lederberg's own interpretation was becoming so convoluted that even he had difficulty believing it. "All of this complexity is a bit hard to take," he wrote to Hayes in April 1952, "and I have often wondered whether I might be on the wrong track altogether."[36]

It was difficult to resolve what was going on during bacterial conjugation when genetic recombination was a rare, one in a million event. The key to understanding it emerged when a new strain of *E. coli* K-12 was found that had a much higher frequency of recombination. Cavalli isolated the strain which he called Hfr in the summer of 1949 and sent it to the Lederbergs to determine whether it was a genetic phenomenon or not. It was indeed. That Hfr strain mated at 1,000 times the frequency of normal crosses.[37]

François Jacob and Elie Wollman conducted an ingenious experiment on the Hfr strain at the Pasteur Institute in 1955, which finally resolved the question over whether the reduction of the genome occurred before or after zygote formation. Using a high-speed blender to interrupt mating of Hfr cultures, they could actually time the progressive entry of different genes from the donor cell into the recipient cell.[38] They also constructed a map based on the time of entry of different genes, and their map agreed with the one Lederberg had constructed.[39]

The time of transfer became the new way of mapping *E. coli* genes, and it effectively closed the controversy over the partial unidirectional transmission of genes from F+ to F-. Only a small segment of the donor chromosome was transmitted to a recipient, as Hayes had proposed, albeit not in the manner that he had advocated. As he quipped in his autobiography, "I had the great advantage of knowing virtually no genetics while Lederberg knew too much!"[40]

Lederberg initially resisted Jacob and Wollman's interpretation of their experiment. But after two or three years, he said, "I exhausted myself trying to disprove it." That was not a matter of obstinacy, in his view; it was just "the way good science works."[41] And even though he conceded that progressive transfer of genes from donor to recipient was the standard event in *E. coli* K-12 crosses, he still maintained that complete fusion of bacterial cells does occur sometimes and whole genomes are transferred from donor to recipient.[42]

Chapter 9 The Extended Genotype

> This review is dedicated to the reconciliation of the attitudes that plasmids are symbiotic organisms, and that they comprise part of the genetic determination of the organic whole. The conflict may arise in part from fixed conceptions of the scope of the organism.
>
> Joshua Lederberg, 1952[1]

Leading evolutionary theorists had come to a consensus about evolutionary processes by mid century: evolution was considered to be a gradual process, fueled by gene mutations and recombination within species. Natural selection acted on minute hereditary variations within populations. Every species has a well-defined boundary, just as the individuals that comprise them have a single well-defined genome.

Lederberg challenged this entire conceptual edifice in the early 1950s when he proposed a far-reaching conception of the importance of hereditary symbiosis. Based on discoveries in his laboratory that viral DNA could become integrated into the genomes of bacteria, and that viruses could pick up and transfer host genes, he argued that biologists' understanding of what constitutes heredity, evolution, and even an organism had to be dramatically modified.

His writings on what he called "infective heredity" culminated in a far-sighted overview titled, "Cell Genetics and Hereditary Symbiosis."[2] It was a masterful piece of scholarship, uniting hereditary symbiosis in bacteria with scattered evidence of its existence in other organisms to construct a new, unified conceptual edifice. It was the first time any geneticist had considered the role of symbiosis in evolution in such scholarly detail.

Symbiosis research developed close to the margins of biology.[3] Nevertheless, there were a few well-known phenomena illustrating the role of symbiosis in evolution: lichens were known to be comprised of fungus and algae, nitrogen-fixing bacteria live in the root nodules of legumes, and mycorrhizal fungi live in association with the roots of forest trees and are important in providing nutrients from the soil. Photosynthetic algae live in the translucent tissue of corals, sea anemones, and flat worms, providing essential carbohydrates for their hosts in exchange for nitrogen. Photosynthetic algae living inside protozoa were already known to reproduce and be inherited from one cell generation to the next.

Yet, microbial symbiosis appeared to be an abnormal phenomenon, rarely achieved, and contradicted what everyone thought were disease-causing effects of infective "germs." Nonetheless, a few microscopists early in the century had suggested that chloroplasts, the seat of photosynthesis in plants, and mitochondria, the seat of respiration in the cells of plants and animals, may have originated eons ago from free-living bacteria that came to live inside the cell.[4] But such views were seen as fanciful conjectures. The United States' premier cell biologist, E. B. Wilson, was more positive about this when he commented in 1925: "To many, no doubt, such speculations may appear too fantastic for present mention in polite biological society; nevertheless it is within the range of possibility that they may some day call for serious consideration."[5]

Lederberg took these speculations about symbiotic mergers seriously, just as he did gene acquisitions from viral infections. In the late 1940s, his close colleague, Tracy Sonneborn, had studied the genetics of a bacterial symbiont called *Kappa* in Paramecium, which was inherited as a genetic entity, and there was a newly studied case of a so-called "viroid" known as *Sigma,* which was inherited through the eggs of *Drosophila* as well. The scope and significance of such phenomena became the subject of controversy in the 1950s. Leading geneticists such as H. J. Muller dismissed such cases, calling them parasites with little significance for genetics and evolution.[6] Chromosomal genes alone, they insisted, constituted the rightful basis of heredity.

Lederberg disagreed. In his view, all hereditary entities had to be considered as part of the hereditary constitution of an organism. In fact, he turned the whole issue about chromosomal genes versus infectious hereditary entities on its head and posited that perhaps many chromosomal genes may also have once arisen as infectious agents. After all, non-virulent forms of pneumococcus could be changed to virulent forms through what appeared likely to be gene acquisition.[7]

The bacterial virus *Lambda* in *E. coli* K-12 could become incorporated into the genetic system of a cell. Its presence remains undetected except when occasionally it becomes active — it then reproduces, kills its host, and is released into the environment. The Lederbergs referred to the relations between *Lambda* and host as an "enduring symbiosis."[8] While there had been a few early suggestions in the 1920s that viruses and their host might live in symbiosis, those claims were either ignored or criticized as contradicting microbiologists' image of "the morbid essence" of the virus.[9] The fertility factor F in *E. coli* K-12 was also shown to be infectious, and it was critical for conjugation.

Then there were the cases of transduction in Salmonella: viruses could pick up multiple host genes and transfer them to new hosts by infection (*Lambda* could do the same, but on a more limited scale). In Lederberg's view, none of these cases of infective heredity could be dismissed as unimportant in evolutionary innovation. The cell-virus association, he reasoned, was "at least theoretically capable of adaptive evolution as a consequence of natural selection and mutation in the virus or the cell component or both."[10]

Lederberg introduced the word "plasmid" as a generic term for any hereditary body, whether it could be transferred by infection or not.[11] In effect, his argument about the plasmid worked in two directions: genetic entities known to be infectious should be treated as part of the hereditary constitution of the organism, and genetic entities not known to be infectious ought to be treated by biologists as if they once might have been.

Lederberg's plasmid concept allowed for agnosticism with regard to the evolutionary origins of intracellular constituents such as chloroplasts and mitochondria.[12] "We should not be too explicit in mistaking possibilities for certainties," he said.[13] The problem was that there were no means available then for determining their origin one way or the other.[14] The same was true of viruses: did viruses originate before cells, or did they arise within cells and then become infectious, as "escaped genes?"[15]

"Hereditary endosymbiosis is probably more prevalent than many biologists are accustomed to believe," he wrote.[16] He saw hereditary symbiosis to be analogous to hybridization insofar as it brought genes from different species together to form a new union.[17] Evolution is not always gradual in accordance with Darwinian theory, but can take leaps through gene acquisitions and symbiotic mergers. He suspected that viruses may also be hidden in animals and remain undetected, just as they are in bacteria. He also pointed to possible incidences of viral transduction in multicellular organisms as evidenced by the production of cancerous tumors.

The Symbiotic Organism

If the concept of heredity and evolution had to be expanded to include infective heredity, so too did our understanding of the boundaries of the organism. In contrast to the one genome-one organism doctrine of geneticists, Lederberg proposed an ecological concept of the organism based on the interdependence of "genetically insufficient" individuals that depend on the biosynthetic products of others. From this perspective, he said, one could see a "graded series of symbioses" from "the co-habitants of a single chromosome" through to plasmids and "to extracellular ecological associations of variable stability and specificity."[18]

Thus, Lederberg challenged the age-old view of the organism as a fixed morphological type, a classification he traced back to Aristotle. Biologists had long abandoned the concept of the species as a fixed type possessing a defining *essence*, in favor of the Darwinian view of species as populations of individuals with different traits upon which natural selection operates. But an essentialist conception of the organism still persisted. Opposing "fixed conceptions of the organism," Lederberg wrote: "The cell or the organism is not readily delimited in the presence of plasmids whose coordination may grade from the plasmagenes to frank parasites."[19] "Hierarchical definitions" should serve the scientist, "rather than the scientist serve an Aristotelian category," he said. "Genetics, symbiotology and virology have a common meeting place within the cell. There is much to be gained by any communication between them which leads to the diffusion of their methodologies and the obliteration of semantic barriers."[20]

In the 1950s, however, very few biologists shared Lederberg's views on the evolutionary importance of infectious heredity and symbiosis. Evolutionists continued to insist that evolution occurred by small gradual changes based on gene mutation and recombination within species. Richard Goldschmidt, one of the well-known rebels of evolution who advocated saltational leaps, or what he called "hopeful monsters," was an exception.[21] He appreciated Lederberg's theoretical discussion of hereditary symbiosis. "As usual among weak humans," he said, "I liked best the parts where my own thinking led me in similar directions."[22]

Leading bacteriologists denied for decades that the kind of gene acquisitions and hybridization shown in the laboratory played an important role in the real bacterial world.[23] But Lederberg's theorizing on symbiosis and infective heredity did have an effect on some, including famed bacteriologist René Dubos. He wrote to Lederberg in 1958 to ask if he knew of any examples of positive adaptive effects of viruses in mammals: "I am amusing myself trying to work up the thesis that the science of microbiology would have taken a very different course if it had used useful effects instead of harmful effects as criteria for selection of microorganisms."[24] Lederberg replied that although he was certain that such adaptive effects existed, there were no techniques available for detecting them.[25]

Dubos would turn toward studying the role of intestinal microbes and the beneficial effects of microbes in the ecology of mammals.[26] He argued that animals evolved in intimate association with microbes, and he suspected that many characteristics of their anatomical and physiological development are manifestations of the response of tissues to that evolutionary relationship.[27] He chided microbiologists for not investigating the positive and beneficial relations between microbes and humans, and instead maintained themselves "as poor cousins in the mansion of pathology." He prophesied that "a new biologic philosophy" would soon develop. "The time has come to supplement the century old philosophy of the germ theory of disease with another chapter concerned with the germ theory of morphogenesis and differentiation."[28] It would be another four decades before that understanding became the foundation of intensive research on what Lederberg called our "microbiomes" in 2001 (Chapter 30).

In the meantime, the importance of infective heredity came to the fore in medicine with outbreaks of multi-antibiotic resistant germs. Mary Barber at Hammersmith Hospital in West London warned of the emergence of antibiotic resistant strains in 1947 when she reported that the widespread use of penicillin and tetracycline had engendered them.[29] Again in 1961, she sounded the alarm of staphylococci that were resistant to multiple drugs and advocated for the controlled use of antibiotics in hospitals "both to prevent the emergence of drug-resistant bacteria and also to avoid rendering patients more susceptible to infection by such bacteria, through elimination of the normal flora."[30]

Lederberg's concept of infective heredity was important for understanding the rapid spread of antibiotic resistance. In 1959, researchers in Japan led by Tsutomu Watanabe reported that multiple drug resistance in strains of Shigella, the intestinal bacteria that causes dysentery, did not arise in a series of discrete steps, but appeared fully developed, and it could easily be transferred between bacterial species.[31] The genes for multiple drug resistance were carried in a plasmid that was transmitted among bacteria by cell contact. Watanabe said it was an example of what Lederberg had called "infective heredity."[32]

Still, as Lederberg had argued in 1952, the generality of the subject of infective heredity or hereditary symbiosis demanded "imaginative speculation checked by the most cautious criticism."[33]

Although he continued to speculate in the 1960s and 1970s that many of our own genes may also have originated as viral infections, such theorizing was far beyond the empirical research of the time. However, the subsequent creation of new molecular methods for evolutionary biology would turn the study of cell origins, microbial evolution, and gene histories into a thoroughly empirical science by the end of the century.[34] Today, it is also known that at least 8% of the genes in our own genomes arose as viruses.[35] Lederberg's idea of an organism as a symbiotic complex is also one of the most important topics in theoretical biology.[36] Gene transfers have been detected every which way throughout the microbiospere, making the pattern of bacterial evolution look more like a web than the iconic branching tree.

Chapter 10

Down Under Immunity

> So I walked right up to this wonderful new theory, and then turned my back on it, and it was Burnet who walked up to it again and didn't turn his back on it.
>
> Joshua Lederberg[1]

Lederberg's research interests had advanced far beyond bacterial genetics *per se* by the late 1950s as he turned to address fundamental questions in mammalian immunology. No one knew how immunity worked: how the immune system produces a specific antibody in response to a specific infection invading the body. When a person gets sick with the flu, for example, their body produces specific antibodies to fight the invading pathogen. That is, a certain immunoglobulin molecule appears in the blood and binds to and inactivates a particular pathogen.

There were two theories about what causes that specific antibody to be produced: one was that the infection induces the formation of that particular antibody, the other was that the infection somehow stimulates the reproduction of that specific antibody-producing cell, among a diversity of such cells already present in the body before the infection.[2] The first theory was the prevailing one in the mid 1950s. The latter is called "clonal selection theory," and it is fundamental to immunology today.

Lederberg made three important contributions to clonal selection theory in the late 1950s. He helped design the key experiments to verify one of its central tenets — that each cell, each lymphocyte, produces only one antibody; he articulated the theory in terms of molecular biology; and he modified the theory in a fundamental way, postulating that antibody diversity was continuously produced by random gene mutations throughout the life of an animal.

It might appear rather odd that Lederberg would turn to study mammalian immunity when he was pioneering the field of bacterial genetics. It happened by accident when he spent three months in Melbourne, Australia, with virologist Sir McFarlane Burnet. Burnet shared the Nobel Prize in 1960 with Peter Medawar for their "discovery of acquired immunological tolerance" in which antigens, initially regarded as foreign, are later accepted as self by the immune system. That discovery was critical for organ transplantations. But Burnet is best known today for developing a model of clonal selection theory at the time when Lederberg was working in his laboratory.

The Lederbergs arrived in Melbourne on Fulbright Foundation fellowships in August 1957. Esther was to co-organize and supervise a practical course in microbial genetics at the University of Melbourne; Josh was to give two lectures a week in microbial genetics at the University and conduct research at the Walter and Eliza Hall Institute for Medical Research where MacFarlane Burnet was director.[3] He had the intention of working on the genetics of the influenza virus.

Burnet, 26 years Lederberg's senior, was well known for his books *Virus as Organism* (1945) and *Principles of Animal Virology* (1955).[4] Lederberg wanted to examine Burnet's data indicating viral recombination, do some experiments, and learn whether specific genes determined virulence. "Burnet was a wizard at the flu," Lederberg recalled, "but he knew nothing whatsoever about genetics. I thought there might be a useful reciprocity in that."[5]

Actually, Burnet had been led to study viral recombination in part because of Lederberg's work showing recombination in *E. coli.* He had written to Lederberg in 1947 saying that he found his research showing recombination in bacteria to be "the most interesting thing on Biology since… Beadle's work in neurospora (sic)".[6] Arriving in Burnet's laboratory ten years later, Lederberg learned how to handle the flu virus in the laboratory and went through basic exercises to do recombination experiments. But his plans took a new direction.

Burnet was in the midst of closing down his research on the flu and turning his attention to immunity. They got into deep discussions about how the immune system produces antibodies when a pathogenic antigen enters the body — whether the infectious antigen induces the formation of a specific antibody, or if that specific antibody is already present in a diverse pool of antibody cells before the infection. Lederberg knew the issues well and had already formed an opinion on it two years prior.

At a symposium on gene-enzyme relations in Detroit in November 1955, Jacques Monod posed the question of whether an inducer such as the sugar lactose in a bacterial medium directed the formation of the specific protein that the bacteria needed to process that sugar. Monod suspected it did. Lederberg disagreed and pointed to evidence from his work with his Japanese student, Tetsuo Iino, on reversible hereditary changes in Salmonella, which indicated that such rapid bacterial changes in response to environmental conditions were not due to directed gene mutations, but rather to regulating gene expression.[7] Monod would come around to this viewpoint and, together with François Jacob in Paris, develop one of the most important models for the regulation of gene expression — how genes could be turned "on" or "off" in bacteria.[8] They would share the Nobel Prize with André Lwoff in 1965.

Lederberg saw antibody formation in response to antigens as a similar kind of problem, in so much as it also would not involve directed mutation. He suspected that specific globulin proteins pre-existed and were then selected for in the presence of an antigen. But he also knew that his views were not widely shared. Even MacFarlane Burnet himself held the opposite view at that time; that is, that the antigen "plays a direct role in moulding the antibody protein."[9]

Soon after Lederberg returned to Madison from the Detroit meeting in the fall of 1955, he learned that at least one other person shared his view about antibody formation. He read an important paper titled "The Natural Selection of Antibody Formation" by Danish immunologist Niels Jerne, published in the November issue of *Proceedings of the National Academy of Sciences USA*.[10] Like Lederberg, Jerne proposed that the antigen would act in such a way as to select from among diverse antibodies already present in the host organism. Jerne would share the Nobel Prize in 1984 "for theories concerning the specificity in development and control of the immune system and the discovery of the principle for production of monoclonal antibodies".

Jerne's paper on antibody formation did not attract much attention at first, but it got Lederberg's attention immediately. He wrote to Jerne explaining the dialogue with Monod in Detroit about enzyme induction:

> Unfortunately, I was not acquainted with your views in time to credit them at a symposium on gene-enzyme relationships which was held

> in Detroit early in November. Monod had gone so far as to imply that even in induced enzyme formation, the inducing substrate acts as a template for "molding" the enzyme — in a rebuttal to this proposal, I was moved like yourself to question the same concept even for antibody formation… At any rate, I was pleased to see at least one explicit hypothesis of the mechanism by which an antigen could "select" one of many synthetic potentialities, and I just wanted to let you know that you have at least one second for your proposal.[11]

Jerne was delighted by Lederberg's response. "I was very happy to have your letter, and as you say 'at least one second for my proposal,'" he replied. "No others have come forth since I published these ideas but I am more content to have you than the whole clan of immunologists."[12]

Still, there were two troubling aspects of Jerne's model. First, it assumed that selection acted on a diverse population of globulin molecules themselves. If so, the globulin molecules would have to be self-replicating, but there were no known self-replicating proteins. The cell itself was the only self-reproducing unit in biology, so it would have to be the unit of selection, not the protein. But that also caused a quandary for Lederberg, because in order for selection to operate on cells to produce a specific antibody, there could only be one kind of antibody produced in each cell. Yet, he suspected that there must be millions of kinds of antibodies — one for every antigen. And if that were the case, the selection theory of immunity seemed to presuppose far too many kinds of pre-existing cells, an unbelievable number.

By the time Lederberg arrived in Melbourne two years later, Burnet, who had also read Jerne's paper, had changed his former view that the antigen directs the formation of a specific antibody to what he would call clonal selection theory. The theory held that a specific antigen would act on a pre-existing group of diverse lymphocytes in such a way as to select for its counter-specific cell, which would be induced to multiply and produce clones for that specific antibody.[13]

Lederberg explained to Burnet that he had already considered and rejected this idea because he thought it would mean that one lymphoid cell made only one antibody specific for one antigen. And if one cell made only one antibody, there would have to be an unbelievably large number of antigen-specific cells to start with.[14] But Lederberg's objection was based on the assumption that there was one specific antibody for every antigen. Burnet rejected that as an untested and thus unwarranted supposition.[15] To be sure, there was a specific anti-flu antibody when you inoculate with flu, and a specific anti-streptococcus when you inoculate with streptococcus, but it was not known if one of these anti-strep antibodies were the same as the anti-flu antibodies. It was entirely possible that there were no more than, say, a thousand antibodies altogether rather than millions, and it was also possible that one cell or lymphocyte made only one kind of antibody. Lederberg immediately saw the error of his thinking.[16]

In October 1957, Burnet sent what is now a famous two-page paper titled "A Modification of Jerne's Theory of Antibody Production using the Concept of Clonal Selection" for quick publication in a relatively obscure journal, the *Australian Journal of Science*.[17] If the concept proved correct, he later said, he would have priority; if wrong, it would go relatively unnoticed.[18]

In the meantime, Lederberg helped design an experiment to test that most troublesome postulate of clonal selection theory — that one lymphoid cell would make only one antibody. He asked Gustav Nossal, Burnet's very bright 26-year-old postdoctoral student, to collaborate.[19] Nossal had also studied bacteriology and had read some of Lederberg's papers in bacterial genetics and knew of his brilliance. He enthusiastically agreed to collaborate with him to test the one-to-one hypothesis.

So Lederberg and Nossal teamed up. They used Salmonella to see what kind of antibody a single lymphoid cell of a rat would make: would it be one specific-antibody per cell or several?[20] Their experimental plan was a technical *tour de force*. First, they immunized rats against two Salmonella strains. Then they isolated single lymphoid cells from those rats and put the cells into little droplets with two Salmonella strains to see if one cell produced antibodies against one or both strains. Lederberg showed Nossal how the experiment could be done using micromanipulation techniques from microbiology.[21] They started the experiments together, but when Lederberg returned to Wisconsin, Nossal continued them by himself while exchanging notes and papers with Lederberg.

The data was unambiguous. The lymphoid cells reacted to only one of the Salmonella strains, not to both. Nossal and Lederberg coauthored their key paper, "Antibody Production of Single Cells," which was published in *Nature* in the spring of 1958. It was the first experimental evidence supporting what would become one of the central tenets of immunity theory: one cell-one antibody.[22]

Lederberg developed clonal selection theory further, modifying it somewhat while reformulating it in terms of molecular genetics.[23] Burnet had good intuition and scientific insight, but he was explicitly averse to molecular biology. Lederberg developed a sharper molecular conception which culminated in his now famous paper, "Genes and Antibodies," published in *Science* in June 1959.[24] Therein he explained what genetic diversification consisted of at the molecular level and proposed a fundamental alteration to the theory. According to Lederberg's model, antibody diversity would continuously increase throughout an individual's life, generated by a high frequency of immunoglobulin gene mutations during lymphocyte proliferation. Burnet adopted this concept in his later formulation of the clonal selection theory of acquired immunity.[25] By the mid 1960s, clonal selection theory based on gene mutations in lymphocyte cells was accepted as fundamental to immunology.[26]

Lederberg's trip to Melbourne had been extraordinarily fruitful in an unexpected way as it led him to address central issues in immunology. But there was still another unanticipated event during the trip that had an astronomical effect on him. Celestial events that he witnessed in Australia and a stop off in India on the way home led him to develop another new field based on the study of cosmic evolution and the search for extraterrestrial life. He called it "exobiology."

Chapter 11

The Andromeda Man

> We hardly want to bring Martian diseases or pests back to Earth as the price of space glory. Since manned flight to the planets is already being seriously discussed, questions of interplanetary hygiene need the gravest consideration.
>
> Joshua Lederberg, 1967[1]

Sputnik was launched on October 4, 1957. Lederberg saw it in the sky over Melbourne. He knew that the space race would soon begin and that military interests and propaganda would overshadow proper scientific exploration, especially in the search for extraterrestrial life and cosmic evolution. He got the US National Academy of Sciences involved in helping to shape space policy by establishing committees with leading scientists throughout the United States, and he moved NASA to launch a new field that he called exobiology.[2]

His deep interest in extraterrestrial life was ignited in November, on the way home from Australia, when he and Esther stopped in Calcutta for a week or so to visit with British biologist, J. B. S. Haldane.[3] A venerable figure in biology, Haldane was well known then for his work on genetics and evolutionary biology, and he had taken an early and friendly interest in Lederberg's work on bacterial genetics.

Haldane had retired from University College London and moved to Calcutta in 1957. Haldane had moved to Calcutta, he claimed, because India shared his socialist dreams.[4] Lederberg suspected that his move to India was actually retirement for the 66 year-old, and his leaving England was much less principled. Haldane had resigned from University College when the university sacked his wife, Helen Spurway, for excessive drinking and for refusing to pay a fine.[5] Lederberg was not interested in Haldane's politics, but he admired his great intellect.

At any rate, it was the day of a lunar eclipse when the Lederbergs arrived in India, and there were religious processions in the crowded streets of Calcutta. Haldane held a dinner for them at the Indian Statistical Institute; he was dressed in Indian attire. With Sputnik orbiting the Earth, discussions over dinner turned to the potential for reckless exploitation of new space technology and how space exploration would be marred by geopolitical competition and used for propaganda rather than science.

It was also the 40th anniversary of the "October Revolution," and Haldane remarked that there could be a second coup "were the Russians to plant a red star on the moon during the eclipse!"[6] That star might be a thermonuclear explosion confirming Soviet military prowess. "At best," as Lederberg recalled decades later, "that was a striking metaphor for the danger that scientific interests would be totally submerged by the international military and propaganda competition."[7]

Lederberg was interested in the possibility of extraterrestrial life and the idea that interplanetary bacterial spores had seeded the earth. It was called the "panspermia" hypothesis and was first proposed by Swedish chemist and Nobelist, Svante Arrhenius, in 1903. Spores could have been launched by intelligent beings or carried aloft from their native planets by air currents and volcanic eruptions, and electrical forces would move them out of the atmosphere.

Such spores might even survive on the moon.[8] Lederberg considered panspermia to be "possibly far fetched" as he wrote to Haldane two years after visiting him in India.[9] But there was a potentially serious obstacle to testing that because rockets would be sent to the moon within a few years and probes sent to Mars and Venus, inadvertently contaminating them with bacterial spores. What would it mean if spores were found on the moon if precautions were not taken to avoid such contamination from Earth missions?

When Lederberg arrived home in Madison in the fall of 1957, he began to read up on rocketry, joined the American Rocket Society, and set out to effect policy on space exploration. He learned that USA's vanguard rocket program was planning to put its first scientific satellite, *Explorer*, in orbit to perform experiments for the International Geophysical Year in 1958.[10] So he started a campaign to try to prevent contamination of the moon. In December 1957, he wrote a memorandum titled "Lunar Biology?" and sent it to a dozen or so scientific notables. "There is at least one important issue that requires advance planning if a unique opportunity is not to be confounded or lost," he wrote. "This is whether or not the meteoritic dust which supposedly covers the surface of the moon contains still viable spores."[11]

He sent a second memo to the president of the National Academy of Sciences explaining his concern about contaminating the moon and recommending that "a committee to study prospects and problems of extra-terrestrial microbiology" be established.[12] The Academy had a mandate to advise government on scientific matters, and Lederberg had just been elected as a member a few months prior.

The response was immediate. The Academy asked the International Council of Science Unions to act, and an International

Committee on the Exploration of Extraterrestrial Space was founded in February 1958.[13] A few months later, the Academy established "The Space Science Board" with Lederberg as one of its founding members. Its mission was to assess the scientific aspects of space exploration, interplanetary probes, manned spaceflight, and the means for searching for extraterrestrial life.

Searching for the cosmic origins of life on earth was not solely a matter of microbes though. Lederberg thought that organic chemicals such as amino acids could have been formed in space and rained down on primitive earth, or else caught a ride on meteorites, thereby jumpstarting the emergence of life. "The problem of organic evolution begins in free space and is coincidental with elementary principles of stellar evolution!!" he wrote in a personal memo in February 1958.[14] That month, he teamed up with physicist Dean Cowie at the Carnegie Institute of Terrestrial Magnetism in Washington to write a paper on cosmic evolution.[15]

They had at a party at a colleague's house in Madison on January 24, 1958. At dinner when Lederberg asked Cowie what he was working on, he said that he was designing an experiment to trap particles of cosmic dust and place them on Petri dishes to look for spores. When Lederberg brought up moon dust, which had been collecting particles for eons, Cowie asked his host for a slide rule and disappeared for several hours, reemerging to say: "you have me beat by 10^{14}."[16] They then agreed to collaborate on a paper.[17] The next few hours were spent on how moon dust could be collected. Someone mentioned that the Soviet Union was sending dogs into space and contamination would certainly happen if one of these landed on the moon. "It looks like we'll have to get busy," Cowie wrote to Lederberg the following week, "or the dog will be splattered all over the moon before we have time to get together."[18]

Their paper titled "Moondust" was published in *Science* four months later.[19] They posited that if the moon were covered in a layer of cosmic dust of great antiquity that had been captured in its gravitational field, and if it had been left undisturbed for eons as astronomers assumed, it would contain "a record of cosmic history as informative with respect to biochemical origins of life as the fossil-bearing sediments on the earth's crust have been in the study of its later evolution."[20] Perhaps Earth and/or other planets served as cradles for the cosmic spreading of life: "Gene flow among planets would alter evolutionary patterns and speed up evolutionary change — short circuit an otherwise tortuous history of evolutionary progress," they said.[21] They called for precautions to prevent contaminating the moon and the planets with microbes carried by interplanetary missiles.[22] It was a timely paper.

President Dwight Eisenhower signed the National Aeronautics Space Act on July 29, 1958, creating NASA that year as a civilian agency, the quintessential Cold War institution to compete with the Soviets.[23] With Lederberg's prompting, it would be quick to embrace cosmic microbiology and studies of the origins of life.[24] Soviet biochemist Alexander Oparin, famed for his theorizing on the origin of life in the 1920s, had already set up meetings on the subject in Moscow in 1957. With fear that the Soviets might lead the research, it was not difficult for Lederberg to convince NASA that the search for extraterrestrial life was cogent to their mission and that inadvertent extraterrestrial contamination had to be prevented.[25]

Lederberg had begun to use the term "exobiology" in meetings of the Space Science Board by the fall of 1959 — "a smaller mouth full than 'the scientific study of extraterrestrial life,'" he said.[26] The following summer, he published a paper in *Science* titled

"Exobiology: Experimental Approaches to Life Beyond the Earth."[27] He considered not only Arrhenius' panspermia but also the possibility of artificial panspermia; that the earth had been seeded by intelligent beings. After all, he said, another century of progressive science could give humans that ability.[28] Although scientists generally were skeptical of the existence of intelligent life elsewhere, he suggested that radio communication with nearby stars ought to be tried.

His main focus was on microbial life. Though one might find non-water based and non-carbon based life forms, he argued, one ought to begin with searching for indicators based on nucleic acids, amino acids, and proteins. It was time to plan space experiments that could be done by remote instrumentation.

Mars and Venus became primary targets for exobiology. Infrared reflection spectrum data indicated that there was an accumulation of hydrocarbons on Mars. "The most plausible interpretation," Lederberg said, "is that Mars is a life-bearing planet." Although the surface of Venus is too hot for life, there might be a more temperate layer at another level. Again, he warned that contamination of planets would "destroy an inestimably valuable opportunity for understanding our own living nature."[29]

The possibility of life on Mars or Venus presented a more perilous risk for space exploration with round trips and even manned space flight, as they saw it, in the not too distant future: back contamination. Lederberg considered the possibility of a new terrifying disease, but he considered ecological risk to be higher.[30] Microbes with unique metabolic pathways could interfere in the normal cycles of carbon and nitrogen upon which agriculture depends, or produce carbon monoxide or nitrous oxide in large amounts. There were many examples of ecological disasters resulting from the introduction of organisms into fresh niches: rabbits and prickly pear in Australia, smallpox in the New World, and syphilis in Europe.[31] "Even the remotest risk" should be mitigated by sterilization procedures, he said.

Lederberg and his colleagues on the Space Science Board considered the possibilities: ultraviolet solar radiation would have a negligible effect in killing microbes that are shielded within a spacecraft.[32] This meant that any space crafts that had entered the atmospheres or surfaces of other planets should not be allowed to land back on Earth until biological studies were carried out to estimate the risk.[33] Such restrictions could be relaxed for the moon because no indigenous life would have evolved there, and any microbes that might have been transferred there on meteorites would also have reached the earth anyway.

Exobiology took off and Lederberg's warnings about interplanetary contamination were heeded. NASA would decontaminate vessels going to the moon and astronauts would be quarantined after returning from space. The USSR would follow suit. In 1960, NASA created its Life Science Office, began funding studies on the origin of life, and embraced the field of exobiology. Richard Young, who started the program on "planetary biology" at NASA, explained it this way decades later: "The switch to Exobiology research turned out to be an important and productive one. It was a new one initially stimulated by a paper written by Dr Joshua Lederberg... These were exciting ideas in exciting times. I got to know Josh well, and was able to fund some laboratory work on his part, in the Life Detection area. He had (and has) one of the best minds I had ever encountered and I had the greatest admiration for his intellect."[34]

Lederberg was elated by the success of his first science policy mission, but he still faced some fierce opposition from colleagues outside of NASA who thought searching for extraterrestrial life was a waste of time and that the fear of interplanetary contamination was simply absurd.[35] He received "a rather hostile" letter from chemist Harold Urey, who not only disagreed with the kinds of precautions against interplanetary contamination that were put in place, but thought that the search for extraterrestrial life was leading NASA down the wrong path.[36]

Urey was someone to be reckoned with. One of the grand old men of science, he had been awarded The Nobel Prize in Chemistry in 1934 for discovering deuterium, and he had played a role in the development of the atomic bomb. In biological circles, he was best known for his theorizing on the origin of life and a stunning experiment by his student Stanley Miller on the origins of amino acids. The experiment was designed to simulate the chemical conditions of a primitive earth atmosphere comprised of ammonia, methane, and hydrogen, which Urey had proposed. Miller exposed the mixture to heat, electric sparks, and water, and within a few days he was able to produce some of the amino acids that proteins are composed of. It was a landmark in the study of prebiotic evolution. He published the results in the spring of 1953, just three weeks after Watson and Crick revealed the double helical structure of DNA.[37]

The "Urey-Miller experiment" was a milestone in studies of the origins of life, but as Lederberg saw it, the conclusion they drew from it — that life arose solely on Earth — was groundless. After all, 90% of the universe is made up of the same few elements that are required for life on earth — chiefly, hydrogen, carbon, nitrogen, and oxygen — so organic compounds should be as common anywhere, even in interstellar space.

The difference between Urey and Lederberg was not simply a matter of data and interpretation. Their scientific styles contrasted dramatically. Lederberg enjoyed speculation while Urey deplored it. Their opposing opinions of Carl Sagan, then a PhD student in astronomy and astrophysics, exemplified those differences. Sagan was known for his speculations on the origins of the cosmos and extraterrestrial life, and he later became famous as a scientific popularizer with his books, both fiction and non-fiction, as well as his PBS television show.[38] While Lederberg considered Sagan to be brilliant and stimulating, Urey had little time for him. "I much prefer a young man like Stanley Miller," he wrote to Lederberg, "who gets some experiments done instead of talking about them."[39]

Lederberg introduced Sagan to NASA, recommended him as an advisor to the National Academy of Science's Space Science Board, and collaborated with him in theorizing on the possibility of life on Mars. "He had a lot of offbeat ideas," Lederberg recalled decades later. "They were always at some level not illogical, and some of them could prove to be right; and I would point that out — the value of listening closely to someone who has that degree of rigor and imagination at the same time."[40]

They had first met when Sagan was a PhD student working at the University of Chicago's astronomical observatory about 70 miles from Madison. Lederberg had a reputation for "brilliance, inaccessibility," and a fierce scientific style, Sagan recalled: "He was an object of some consternation and fear. Post doctorates in biology were afraid to present papers lest he be in the audience and demolish their thesis with two questions. Then, one day, he called me up out of the blue and said he wanted to see me — said he was interested in extraterrestrial life. I was immensely flattered."[41]

Sagan saw Lederberg not as someone to be feared, but rather as "a dry, stimulating, totally unfettered man" who was "willing to carry his ideas to their logical consequence, even though the prevailing wisdom says it's silly."[42] They took to each other immediately and enjoyed bouncing ideas off each other. "We didn't either of us have to finish sentences," Sagan recalled. "We could leapfrog through arguments — an efficient way of talking."[43]

Lederberg not only supported Sagan's imaginative ideas, he also supported the speculations of Sagan's wife, Lynn Sagan (later Margulis). She was a Master's student in cell biology at the University of Wisconsin when they first met.[44] NASA would fund Margulis for decades as she became one of the most well-known biologists in the world, famous for championing the role of symbiosis as a source of evolutionary innovation.

Lederberg, as it will be recalled, had considered such ideas in a masterful overview on hereditary symbiosis in 1952 (Chapter 10). Margulis advanced those arguments for symbiosis in cell evolution and Lederberg encouraged her.[45] In 1966, when her now classic paper on symbiosis and the origin of cells was rejected with harsh comments from the editor of *Science,* Philip Abelson, Lederberg wrote to her, "Don't be put off by Abelson's response; he was being more cordial than usual… I had a hardly different comment from him when I submitted the enclosed." That was another manuscript on exobiology.[46]

Abelson, indeed, had no time for exobiology. He was more than skeptical that life, as we know it, would be found on the moon or on another planet. He gave a talk on this at the National Academy of Sciences meeting in Washington in December 1960.[47] He did not mince his words: "In looking for life on Mars we could establish for ourselves the reputation of being the greatest Simple Simons of all time."[48] Lederberg thought Abelson's perspective was simply shallow.[49]

The debate over the risk of back contamination would heat up after Lederberg wrote an article in *The Washington Post* on February 26, 1967 titled, "Are There Bugs on Mars?" and warned of a space craft accidently carrying pathogens back to Earth: "Since manned flight to the planets is already being seriously discussed, questions of interplanetary hygiene need the gravest consideration."[50] This time, he and his collaborators became caught up in debates in letters to *Science*.[51]

But before the decade was out, another issue arose, not over the risks of interplanetary contamination *per se*, but over Lederberg's own image as a scientist when Michael Crichton's first best-selling thriller, *The Andromeda Strain*, appeared in 1969. In this tale, a satellite, which was designed to capture upper-atmosphere microbes for bio-weapons exploitation, returns to earth carrying a deadly extraterrestrial microbe that was acquired when a meteorite crashed into it. The microbe, codenamed the Andromeda strain, kills everyone in the Arizona town where it lands. One of the main scientists in the story is a man named Jeremy Stone, head of the bacteriology department at Stanford, who had received a Nobel Prize in medicine in 1961.

It was obvious to Lederberg and his friends that Stone was modeled after him. He had moved to Stanford in 1959 and was awarded the Nobel Prize a few months earlier. *The Andromeda Strain*'s publisher, Knopf, claimed in its promotions that the book had "documentary verisimilitude." Lederberg asked for the book to be properly fictionalized because it was inaccurate and cast his character in "an uncomplimentary light."[52] When he also learned that a film version of the book was planned at Universal Studios, he wrote to its producer and director to request that "reasonable and prudent precaution" be taken to minimize confusion between him and any of the characters.[53]

As it turned out, the assumption underlying Lederberg's proposal that the moon's surface would be a likely place to look for spores because it was undisturbed for eons was incorrect. The moon's surface, as well as the surfaces of planets, was thoroughly reworked by meteorite impacts. Life on Mars, however, remained a good possibility.[54] In 1960, Lederberg established a NASA-based instrumentation laboratory to probe Mars using automatized technology (Chapter 16). Lederberg's suggestion that organic molecules might be formed on interstellar grains found support beginning in 1984.[55]

When astrochemist Mayo Greenberg published an article with experimental evidence for it, Lederberg wrote to him, "My interest in interstellar grains has not diminished and of course, some facts from contemporary investigation are a welcome successor to speculation."[56] Greenberg replied: "How prescient you were!"[57]

Chapter 12

Berkeley Debacle

> We hear rumors of your offers from the West. We hope you will yield. We would all be very happy to have you closer.
>
> Max Delbrück to Joshua Lederberg, February 25, 1957[1]

Lederberg was becoming restless after ten years at Madison. If some members of the genetics department at the University of Wisconsin had had concerns about hiring him in 1947 (Chapter 6), a decade later the problem was reversed — they now had concerns that he would leave. The abrasive personality that his colleagues experienced when he was a graduate student had softened. His influence on campus was important, not only for the development of the genetics department, but for biology as a whole. As the Chair of the genetics department, Alexander Brink, recalled, "he had many of the characteristics of a genius and was just bursting with mental and physical energy."[2]

Lederberg's closest colleague at Wisconsin, James Crow, a renowned mathematical geneticist, had the same impression. He and Lederberg had adjoining offices and they conversed almost every day. Before there was Google, Crow commented decades later, they had Lederberg. He seemed to know everything, and what he did not know he quickly found out, and his memory was formidable: "I once showed him how to derive the Kosambi gene-mapping function. One evening in my office, a year or more later, he dropped in for a casual visit, as he often did. I had forgotten how the derivation went, so he immediately straightened me out. This suggested a strategy: tell him something and have him remember it for me. I did not actually exploit this, but it happened nevertheless."[3]

Lederberg had become known as one of the most productive biologists in the country as he and his collaborators put out one outstanding piece of work after another in the 1950s from his little laboratory at Wisconsin. Almost every month it seemed, there was something new: *Lambda* phage, the fertility factor F, transduction, and occasionally new techniques like penicillin selection and replica plating. He had developed one of the central theories in mammalian immunity and theorized on hereditary symbiosis in evolution. He championed the biological opportunities and responsibilities of space research beginning in 1957 and got NASA interested in the field he called exobiology.

At Wisconsin he rose up the ranks quickly to full professor by 1954 at the age of 29. The American Society for Microbiology awarded him the prestigious Eli Lilly prize in 1953, the first time it had ever been awarded to a geneticist. As one of his colleagues humorously wrote to him: "Congratulations on receiving the Eli Lilly Award. I was beginning to fear that you'd be always a bridesmaid, but then sex is always controversial and slow to be accepted especially among bacteriologists."[4]

So it was hardly surprising that Lederberg had begun to receive several job offers from elsewhere. The Albert Einstein College of Medicine invited him to establish a bacteriology department before it opened its doors in 1955.[5] He rejected the offer as he was interested in medical genetics, not just medical microbiology, and he did not want to return to live in New York City. He enjoyed the bucolic atmosphere of Madison, the collegial life at the university, and the ease of getting to work — a relief after his very hectic and frenzied life in New York City.[6]

Nevertheless, he was getting restless — and as the attractive offers came pouring in, his colleagues at Madison made every move to keep him. In fact, the university created a department of medical genetics in the medical faculty for him in the spring of 1957 to that end. He had written a proposal for the department in December 1955.[7] When the news broke that Lederberg might leave the university, a department was quickly formed with two faculty members and himself as Chair.[8] It was one of the first medical genetics departments in the country, but it would not be enough to keep Lederberg in Madison.

Some of the most prestigious universities in the country wanted him, but Lederberg entertained only a few.[9] Harvard made an offer in February 1957.[10] The Biology Department wanted to build up its genetics program. Lederberg would be part of a trio with Paul Levine and James Watson of double helix fame, who had been hired a year earlier.

But Lederberg was circumspect: the wheels of change seemed to move too slowly at Harvard. It had made gestures to offer him an appointment five years earlier, but nothing had come of it.[11] This time though, the university meant business and he received several enticing letters. The Chair of the biology department wrote to him about the good quality of the students and about Cambridge life, how the Boston Symphony plays a short series of concerts right on campus, and so on.[12] Evolutionist Ernst Mayr, who was hired by Harvard in 1953, explained how he enjoyed the contact with outstanding scholars in all fields and found the university scientifically stimulating. "Your coming here would give genetics and microbiology a tremendous boost," Mayr wrote, "and I am sure you would not regret it."[13]

Still, there was animosity toward genetics among some non-experimental biologists at Harvard, which became apparent when a hullabaloo erupted in the biology department over the promotion of Paul "Levine" to Associate Professor in 1957.[14] Biochemist Paul Doty wrote to Lederberg about how much he was needed to "bring about a renaissance in genetics at Harvard."[15]

Watson explained to Lederberg that he had no regrets after being at Harvard for a year. He admitted that the biology department had not "been up to decent standards" and "muddled along," but explained that things were changing now that the administration was out "to correct this mess and hire the best people."[16]

But their efforts proved futile. Lederberg had his eyes on the University of California, Berkeley, where he had spent a few weeks in the summer of 1950 as a visiting professor. Lederberg liked everything about California, especially the San Francisco Bay area. It was "one of the most exhilarating parts of the country," he wrote to a friend that August.[17] He and Esther stayed way up in the hills, and it planted a seed in his mind: this might be a place he wanted to live someday.[18]

Several of the country's leaders in genetics were at the University of California at Berkeley then. The bacteriology department was also one of the best in the country. Venerable microbiologist C. B. Van Niel had taught a summer course in California since 1938. It had been important in training many of the scientists who became leaders in microbial genetics, including both Lederberg's wife Esther and his brother Seymour.[19]

But Lederberg's closest association in the summer of 1950 was with Roger Stanier, who had invited him to Berkeley to give lectures in the bacteriology department. Lederberg was taken with Stanier's breadth of interests and insights into microbiology and biochemistry. At first, he wanted to see if he could get Stanier to move to Madison, and he thought he had the opportunity to do so because of a standoff between Stanier and the University of California over McCarthyism.

Those were testy days at Berkeley when the Lederbergs arrived after a five-week road trip from Madison in the summer of 1950. McCarthyism had become a hot issue for academics there and elsewhere. They were supposed to sign an anti-communist loyalty oath. Lederberg had no time for fascists like McCarthy any more than he did for communists, who he believed were a real threat. He had, from a very young age, deplored the people who couldn't see through the Molotov-Ribbentrop Pact, the non-aggression treaty of 1939 between Nazi Germany and the Soviet Union. For Lederberg, that was the dividing line. For anyone to call themselves "an 'anti-fascist' and still adhere to the party after that point seemed an absolute absurdity."[20]

Still, Lederberg had no sympathy for "the stupid 'loyalty oath,'" especially because it was demanded of professors. Why did university faculty have to prove their loyalty to the United States? Those who did not sign it were "fired, not for any suspicion they were communists, but for discipline!"[21] Ironically, because Wisconsin was McCarthy's home state, the University of Wisconsin did not require its faculty to sign the oath. Lederberg stood up for civil liberties and academic freedom.[22] As he saw it, the university faculty was "in the best position to judge its own members in the interest of the university and of the nation."[23]

Stanier also opposed the loyalty oath. When the Regents of the University of California demanded that all university employees sign the oath, he agreed to sign it, but promised to resign if any members were fired or suspended for refusing to do so. When that did happen, Stanier intended to follow through on his promise, which presented an opportunity for the

University of Wisconsin to hire him.[24] But the president of the University of California resolved matters so that Stanier could remain at Berkeley. Lederberg then turned back to the idea of moving to California, and that desire was reinforced when he went to Berkeley again in 1953 to receive the Eli Lilly Award in microbiology.[25]

An opportunity seemed to present itself when a position in genetics opened up at Berkeley in January 1957. Still, there were two issues. One was that the position in genetics was in the agricultural college at Berkeley. Lederberg wanted to develop medical genetics like he was doing in Wisconsin by that time.[26] The other pressing problem was obtaining a position for Esther. She was essential to her husband's laboratory and career, effectively the laboratory "CEO" as well as research collaborator, and her role allowed him to develop and maintain a broad diversity of interests outside the laboratory.

The problem was "anti-nepotism" rules. During the 1950s, anti-nepotism adversely affected many women scientists who had received their PhDs and were married to faculty members. In some cases, women were able to work in their husbands' laboratory but could not be employed by the university. In other cases, they were not permitted to teach if they worked in the same department as their husbands.[27] Wisconsin overcame the nepotism problem because Esther was paid through an army contract from 1950 onward.[28] Her position at the University of Wisconsin was "Project Associate in Genetics," with a rank and pay about equivalent to a starting assistant professor.[29]

Discussions to resolve the nepotism obstacle at Berkeley would continue for some fifteen months. The decision to hire Esther had to be made at the highest branches of administration under the University of California system. Several ideas were aired. One was that Esther could receive a grant to support her work in her husband's laboratory, if that grant were given to another department. Josh was appalled by the situation. He wanted a proper position for Esther in her own right as a research associate in his laboratory. "It seems to me utterly silly," he wrote to the chair of genetics on January 15, 1957,

> that a regulation of this kind — which is designed to forfend real abuses in other contexts to be sure — should frustrate and waste the professional lives of trained and competent women who happen to be married. For my own part, it would be impossible for me to maintain a diversity of interests and personnel working smoothly without the help of such an associate.
>
> I could also point to our joint publications, which are an imperfect measure of the role of her collaboration.[30]

Lederberg also explained that "her work on the genetics of lysogenicity has had a decisive impact on the phage world, and the *Lambda* system is now one of the most widely studied in that field." She also "had a primary part" in the work on the fertility factor and "was an indispensable collaborator, to say the least," on other papers. "She has also had much to do with the tooling-up for many other studies where she does not appear as an author."[31]

In May 1957, the president of the University of California, Robert Sproul, approved another proposal: Esther would be employed in another department on regular university funds as a "research microbiologist in the non-faculty professional research personnel series" and would be free to work in the genetics department on problems of her own choosing.[32] It wasn't perfect, but Esther was "quite gratified at the proposal."[33]

So it was settled, and by mid-January 1958, big plans were afoot at Berkeley. The department of genetics was going to be moved out of the agricultural college and a new $220,000 microbial genetics laboratory created for Lederberg.[34] Everything finally seemed to be on course when suddenly two months later and for unexplained reasons, everything fell apart. President Sproul decided not to go through with the offer. No one saw it coming and everyone was shocked.[35,36]

It had all been a complete waste of time, and it ended in a bit of a mystery as to what really happened. The official statement in the archives was that the president's office did not want to set a precedent by hiring the Lederbergs in the same department.[37] But that explanation seemed to be erroneous. After all, president Sproul had agreed on Esther's being hired in a different department from her husband. As Lederberg came to learn years later, the whole issue seemed to be due to an administrative power struggle between outgoing and incoming presidents of the University of California system: Sproul was retiring and was to be replaced by Charles Kerr, who had been Chancellor of Berkeley.

Kerr explained the events to Lederberg at a Carnegie Institution dinner in New York in November 1978. He made a point of going over to Lederberg to inform him that he had a vivid and distasteful recollection of the events. As he explained it, president Sproul, who had previously been Chancellor of Berkeley for 22 years, did not want his administration at Berkeley to be followed by any greater accomplishments.[38]

Ironically, by the spring of 1958, Lederberg was probably gratified that things had not worked out at Berkeley, because by that time, and after all the Berkeley delays, he was in the midst of discussions with Stanford.

Chapter 13

How the West was Won

It seems to me we need some coordinated actions to get Lederberg and Kornberg or we'll lose them both… Can we lay out a plan for capturing these two gentlemen? What a day that would be for Stanford!

Avram Goldstein to William Steere, February 4, 1957[1]

Who's going to be left in the Midwest?

Lederberg to Arthur Kornberg, January 27, 1958[2]

The *contretemps* at Berkeley was over by mid-March 1958, but fortunately, Lederberg had kept several balls in the air. In fact, he was already in the midst of finalizing negotiations with the medical school at Stanford University — the perfect fit, as it turned out. Actually, he had been offered a position in the biology department of Stanford in 1956; he and Esther had gone there for a visit during the winter of 1957, but he was not keen on the position or the university.[3]

At the time, Stanford was a sleepy place with a kind of country club atmosphere rather than a hot bed of new ideas and innovation like Berkeley. But Lederberg's attitude toward Stanford changed dramatically when he learned that it was set to create a new and innovative medical school. The old medical school was in San Francisco some distance away, while the new medical school was to be constructed on the main Palo Alto campus. Moreover, it would have a dramatically new and different ethos than the traditional medical trade schools focused on rote training that

Lederberg had deplored as a student at Columbia College before he dropped out in 1947. First, it would be research-oriented and focus on graduate as well as undergraduate education.[4] Second, it would encourage interaction among various disciplines. One of the main reasons for moving the medical school to the main campus was to interweave medical science with the physical and biological sciences, as well as with the social sciences such as psychology and sociology.

So Lederberg began to look at Stanford in a whole new way by the spring of 1957. Avram Goldstein, the chair of Stanford's new pharmacology department, did much of the work in getting Lederberg to reconsider Stanford where he himself had moved to from Harvard in 1955. Goldstein was a noted expert on addiction and one of the discoverers of endorphins. He had invited Lederberg to give a lecture in 1952, and they subsequently corresponded about bacterial genetic methods and microbial resistance to drugs.[5]

The development of the new medical school and its move to Palo Alto had lagged during Goldstein's first year at Stanford, but by January 1957, Goldstein wrote to Lederberg that there were now "tremendous possibilities for strengthening the medical school." But he had also heard "disquieting rumors" that Berkeley was "dangling bait" before Lederberg.[6]

Goldstein had some better "bait" of his own to dangle before Lederberg: Arthur Kornberg was thinking of moving to the medical school at Stanford. Kornberg was one of the best biochemists in the country, known especially for his research on the enzymatic machinery by which DNA is replicated. When Watson and Crick published the structure of DNA as a two-stranded double helix, they saw how one strand — half the DNA ladder — might act as a template for copying a new strand. They made no mention of enzymes and thought that DNA was somehow a self-replicating molecule. Actually, one can still hear often that DNA is self-replicating. But it isn't; in 1956, Kornberg and coworkers identified and partly purified an enzyme, DNA polymerase in *E. coli,* and showed how it catalyzes the production of new DNA strands.[7]

As Lederberg was interested in developing molecular biology, an opportunity to work in close association with Kornberg could not be beaten. Wisconsin had great biochemists, but none were interested in the biochemistry of DNA, protein synthesis, or gene action.[8] Kornberg was seven years older, but their early histories were similar. Kornberg was also a second-generation New Yorker of Eastern European Jewish heritage. His father had worked as a sewing machine operator in the sweatshops of the Lower East Side of Manhattan for almost 30 years before opening up a hardware shop in Brooklyn where young Arthur worked from the age of nine. Academic achievement fostered by the public schools was for him the path to upward mobility, no less than it was for Lederberg.[9] Kornberg's wife Sylvie was also a biochemist, and like the Lederbergs, the Kornbergs frequently collaborated.

Goldstein wanted to move quickly to get offers to Lederberg and Kornberg. "I am glad to do what I can," he wrote to the Dean of Graduate Division on February 4, 1957. "Re: Lederberg, could we explore the Medical Genetics angle, which would greatly benefit the med school? [...] Also would a fast approach to Kornberg be instrumental in getting Lederberg? I think so."[10] Things got rolling decisively.

In June 1957, Kornberg was offered the position of "Professor and Executive" of a new biochemistry department. He was bringing with him his entire biochemistry department from Washington University in St Louis, a group of seven faculty members.[11] They were going to move *en masse* when the new medical research buildings were completed in 1959. Goldstein relayed the news to Lederberg:

> When we talked at our house you said, "If Kornberg comes to Stanford I'd be on the next plane." It appears now that he will, but what happened to the plane schedule? [...] Is there anything I could possibly do that would move you here... A department of medical genetics in the Med school, a joint appointment with biology? We are moving fast in the right direction and nothing should be impossible. Or are you and Esther now wedded to the cold winters for life?[12]

That message certainly got Lederberg's attention.

The effort to hire Lederberg came from multiple sources. Henry Kaplan, who had pioneered radiation therapy for cancer patients and was Chair of the radiology department in the medical school, was also involved. In June 1957, he was at a party in St Louis celebrating Kornberg's department's imminent move to Stanford. Just as he and Kornberg walked in the door, the phone rang. It was Lederberg wanting to speak to Kornberg. Kaplan recalled: "He wanted to know what was going on at Stanford. He said he had been out there last year, looking into a job offer in the biology department and turned them down, because it seemed like the same old sleepy place as ever."[13] But he had just heard that Kornberg was moving there and he wanted to find out what was happening to get in on the excitement. "We took turns on the phone," Kaplan said, "and talked about the new and improved curriculum, outstanding students and so on, and Josh said, 'Gee, that sounds wonderful! I'd be very interested.'"

Kaplan flew back to Stanford that night, went in to see the dean of the medical school, Robert Alway, the next morning, and "demanded" that he forthwith create a new department of genetics with Lederberg as chairman: "Alway had to dig down into some temporary dean's funds in order to get it off the ground. Alway, to his ever-lasting credit, was never afraid of people who were smarter than he was. And he was not afraid to do battle with the university administration if the issue seemed important enough. That's how it happened."[14]

Well, that's one version of how it happened. Goldstein had been working with administrators for months to get Lederberg. Kornberg also got involved. He wrote to Lederberg on August 19, 1957 to stress the advantages of Stanford over Berkeley while explaining how the faculty of medicine would not be based on the awful trade school model, and that administrative headaches would be minimized.[15]

Lederberg explained his situation to Kornberg in January 1958.[16] He had an offer on the table with Berkeley but it had not yet been finalized, and he was not even sure if he could actually leave Wisconsin because by that time the university had established a small genetics department in the medical school consisting of himself and one other person, and they were in the throes of hiring another. If a third person were hired, he thought the department of medical genetics could survive without him. Genetics counseling was at such a primitive stage that they decided to "avoid it like sin." Once they "train more physicians who know what a chromosome is," they could offer to practitioners information to help them cope with any particular situation.[17]

Lederberg then made two suggestions to Kornberg. The first was to establish a genetics department in Stanford's medical school. But he was leery of being chair of a large department with all the administrative duties that would entail. It would have to be a small one. The second idea was that he and Kornberg establish a single "Department of Biochemistry and Genetics." "After all," he wrote, "Art, you are to my mind Stanford's most valuable asset. It is perfectly plain how uniquely your talents and interests complement mine — my most cogent dissatisfaction with Wisconsin is that I could not succeed in working out a long lasting biochemical collaboration, which I badly need."[18]

Kornberg had nothing but admiration for Lederberg, but he had no interest in being in a department with him. Their styles were worlds apart. Kornberg's career seldom wandered far from the laboratory bench. He was an enzyme hunter, and he said he never "found a dull one." Lederberg's interests had already broadened to include various aspects of bacterial genetics, symbiosis, immunology, medical policy, mutagenesis, and cancer. Exobiology was also on the horizon. "Lederberg really wanted to join my department," Kornberg recalled decades later. "I knew him; he is a genius, but he'd be unable to focus and to operate within a small family group like ours, and so, I was instrumental in establishing a department of genetics of which he would be chairman."[19]

With the support of Stanford's president, provost, and the dean of the medical school, Kornberg went to Wisconsin to meet with Lederberg as an "ambassador" in order to facilitate hiring him. He wrote a report on February 5, 1958 explaining what would be an irresistible offer: "Creating a Department of Medical Genetics with Lederberg as its head," Kornberg concluded, "would dignify its importance to students and faculty and would give Stanford enormous prestige."[20]

Stanford immediately worked up the offer.[21] "Personnel, space, curriculum hours and other essential considerations seem scarcely insurmountable to us," Dean Robert Alway wrote to Lederberg on April 7, 1958, "So, would you and your wife consider an early visit to Stanford?"[22] The Lederbergs flew out the next month.

In the meantime, Lederberg sent Alway a description of what he had in mind for the department. It would offer a teaching program of genetics to medical graduate and postgraduate students and maintain its own research program.[23] He wanted there to be broad training on central questions in genetics, rather than simply focusing on human genetics to the exclusion of experimental work. Genetics research, he said, would be based on "microbes, mice and man." The syllabus aimed to have a balance between the fundamentals of genetics and practical applications in diagnosis and counseling. After students were taught the principles of genetics, they would then be exposed to the genetic aspects of various diseases. The department would also encourage counseling in collaboration with clinical departments, especially pediatrics.[24]

Lederberg wanted a little "pocket-sized" department with a few handpicked faculty members in addition to research assistants and postdoctoral fellows. He also wanted his department to be in the main medical research building, as close as possible to Kornberg and his crew in the biochemistry department, so as to have a day-to-day working relationship with them. Kornberg wanted the same and gave up some of his space in the corner of his department to accommodate some of Lederberg's requirements.[25] He needed about 2,500 square feet of space — three times the size of his lab at Wisconsin — with remodeling costs that would exceed $100,000.

As it turned out, funds were not hard to find. Stanford had just come into a windfall. In April, the Ford Foundation awarded it 2 million dollars, and a local foundation gave the medical school another million.[26] Funds for Lederberg's department were approved, and a formal offer was made on July 1.[27] His salary would be $16,000 per year.[28] There were no last minute setbacks.

Esther would be hired as a "Research Associate" in the genetics department. She would not be a paid university employee, but would be "eligible for appointment to a non-tenure position payable from grant funds."[29]

Lederberg was careful in these negotiations not to offend the University of Wisconsin, which had been so supportive of him and had hurried to establish a little medical genetics department to try and keep him. But as he explained in his letter of resignation to the university's president on July 19, 1958, the special opportunity at Stanford to work in close quarters with Kornberg was a major factor in his decision to move.[30]

Lederberg took up his position at Stanford on February 1, 1959. Kornberg and his group arrived eight months later, and the medical school would be transformed from a mediocre clinical school to a powerhouse for medical research. As Kaplan put it, "Suddenly Stanford was catapulted literally from a second or third rate clinical school into some kind of mysterious but very exciting place that students thought of in the context of Harvard."[31] Lederberg would be at the heart of that renaissance. Four months before his arrival, people at Stanford learned that they had hired not only a renowned professor, but also a Nobel Laureate.

Chapter 14

Nobel Politics

> It's a wise man who can choose his own parents; wiser still if he can choose, or be chosen by, his associates. The catastrophic distinction of the Nobel Award makes me feel this more keenly than ever.
>
> Joshua Lederberg to former students and collaborators, October 30, 1958[1]

It was a frantic Autumn. The Lederbergs had spent four months away in Australia, India, and Europe. They had made their way past the Berkeley debacle and were preparing to move to Stanford early in the new year. Then, on Sunday, October 26, things got really complicated. Josh went to the lab alone around 11 am to work on two grant applications. At around 11:30, he received a call from a Swedish reporter who leaked the news that he, along with George Beadle and Edward Tatum, were to be co-recipients of the Nobel Prize in medicine. The announcement was going to be made on Thursday. He did not keep a diary in those days, but he decided to take notes on this event.

Lederberg was incredulous. He was not surprised that Beadle should be honored this way for his pioneering work on biochemical genetics, and he considered it a "perceptive courtesy for Tatum," Beadle's collaborator, to share it. But he was astonished that he was included. He was far too young. The Prize was for "the venerable and veterans" of science. "I guess I just don't believe in memorializing the live and kicking," he wrote. "On the whole I'm a little afraid the fuss and bother more than outweigh the egotistic satisfactions, the cash and the prestige factors that might help in getting my lab going. Perhaps I'm exaggerating the fuss… I am rather nervously awaiting the AP bulletin to be picked up locally as I'm sure I'll have no peace after that!"[2]

He considered not accepting it. After all, the Prize presented a distorted image of science in so much as it celebrates heroic discoverers, when science is in fact a collaborative activity. No matter who the hero is, that individual, as he put it, "must parasitize dozens of people who do not directly share recognition but may get secondary benefits." His own research involved talented students and collaborators such as Norton Zinder and above all, his wife, Esther. The Nobel Prize was a farce, so why go through with the pretense? Anyway, he suspected that Nobel had only established the prize to appease his conscience for inventing dynamite.

But rejecting the prize would only cause more notoriety. "I do feel as much as ever that the nonsense ought to be abolished," Lederberg wrote "but I don't have the courage to meet it head on and I'm afraid it would raise even more fuss and perhaps affront Ed and Beadle in a rather nasty way." He was trapped; there was no way out, and he knew it. There was also no denying that the Prize would be great for his research program when he moved to Stanford. The reporter's leak that Sunday morning was not wholly reliable, so he kept his head low for a few days. Then another story appeared in the newspapers, and the official news came in a Western Union Telegram from the Rector of the Karolinska Institute in Stockholm.[3]

There was another extraordinary aspect to Lederberg's winning the prize in addition to his youth. It was awarded to him in the first year that he was nominated.[4] Beadle, for example, had been nominated 14 times since 1948. Tatum had been nominated on six occasions since that time. Albert Einstein was nominated for the Prize in Physics 62 times over a period of 11 years beginning in 1910.

Lederberg's nomination came in just two steps. First, the Dean of Harvard's Medical School, George Berry, nominated him in January 1958. Berry's colleague, Bernard Davis, wrote much of the nomination letter on behalf of "the Harvard group." Davis knew as well as anyone how Lederberg and his collaborators opened up the field of bacterial genetics with one fundamental paper after another. Berry and Davis also tried to recruit Lederberg for a new Medical Genetics Chair in the Medical school, for which they had just acquired funds in July 1958.[5] But unfortunately for them, he had just came to Stanford's medical school two weeks earlier.

The Nobel Prize was ostensibly aimed at individuals, but it was also for the field of science. And as Berry stated in his letter to the Nobel Committee in Stockholm, there was "no area of biology in which fundamental knowledge is being more effectively acquired" than microbial genetics.[6] He explained the reasons why: 1) DNA, previously considered to have little diversity, was shown to be the basis of the gene. 2) Gene function was based on the systematic study of microbial mutants that blocked specific metabolic pathways. 3) Bacteria, previously considered to have inheritance only by binary fissions, have been shown to undergo recombination by cellular contact and, in certain instances, also by transduction when viruses passively carry DNA fragments. 4) Mutation and recombination was also shown to occur among viruses. 5) Lysogeny was understood to involve the incorporation of a viral DNA on a particular site in bacterial DNA.

Berry nominated Lederberg and Beadle, along with Max Delbrück, Alfred Hershey, and André Lwoff. These awards, he said, "would form a suitable background for the inevitable later consideration of the work of Watson and Crick on the structure of DNA and that of Kornberg on its enzymatic synthesis."[7] As it turned out, all of them would eventually be awarded the Prize.

Lederberg's friend, George Klein, a member of the Nobel Prize committee, took Berry's nomination letter further to elaborate on why Lederberg should be awarded the Prize. A brilliant cancer researcher at the Karolinska Institute, Klein had immigrated from war-torn Hungary 10 years earlier, having narrowly escaped the gas chambers at Auschwitz.[8] He married Eva Fischer who, like him, had been a medical student at the University of Budapest before she and her family went into hiding. The couple left Hungary to live in Sweden in 1947 without knowing a word of Swedish, and completed their medical degrees at the Karolinska Institute. Both became professors there, and a new Department of Tumor Biology was later created with George Klein as its head. The Kleins often collaborated together as did many other married couples in the tumor biology department. "At one point we had seven married couples working in the lab at the same time, surely a world record," they said.[9]

George Klein had first learned of microbial genetics at the International Congress of Genetics held in Stockholm in 1948. Knowledge about cancer was rudimentary then and Klein was interested to know if cancer cells, which were resistant to growth regulation, could be compared to some mutations in microbes.[10] Lederberg had suggested that very idea in a note published in *Science* two years earlier when he was a 21 year-old medical student. He and Klein began to correspond and exchange papers in the early 1950s as Klein started using microbial genetics methods to study some aspects of tumor cell populations.

They finally met for the first time at a tumor conference in New York in May 1955 and they took to each other immediately, spending the better part of two days together.[11] They met again when the International Congress of Microbiology was held in Stockholm in August 1958. Lederberg chaired the microbial genetics session.[12]

Lederberg knew nothing then about Klein's role in nominating him for the Nobel Prize six months earlier. After receiving Berry's nomination of Lederberg for the Nobel Prize, Klein dove into Lederberg's papers as a Nobel committee member and wrote a comprehensive 20-page overview of his work in Swedish.[13] Although he made no mention of the nomination to Lederberg, he did hint at what might be going on in his correspondence. "I now definitely expect to see you again before the end of this year," he wrote to Lederberg on September 23, 1958. "I cannot say more about this now but will let you know about it later."[14]

Lederberg made nothing of it — until news of the Nobel Prize came from reporters. "George, I hardly know what to say!" he wrote on October 29:

> I am full of appreciation to you for your inevitable part in proposing the Nobel award, and for the fuss and hard work it must have meant for you. That our personal friendship can have played no role whatever in a matter of this kind does not diminish its additional warmth on the occasion, and I am sure also that we are equally delighted to accomplish this recognition for genetics and especially for genetics in relation to medicine.[15]

Klein was overjoyed that his nomination had gone so well. "I shall not deny that I had a certain amount of hard work to do in reading through your monumental production and trying not only to understand it but also to interpret it to colleagues in all fields of medicine and in beautiful Swedish," he replied.

> However, I did seldom have a job with more pleasure and satisfaction… You can be absolutely sure that our personal friendship played no role whatever in this; I can safely say that seldom has a candidate with so many important achievements been considered as in this case.[16]

He then explained the politics of the Prize.[17] First, in accordance with Nobel's will, there could only be three Prize winners in each category: physics, chemistry, and physiology or medicine. Second, although the Prize was really awarded for a body of work, the attribution had to be to something at least resembling a single "discovery." The result was rather odd: Beadle and Tatum shared one half "for their discovery that genes act by regulating definite chemical events," and Lederberg the other half for "his discoveries concerning genetic recombination and the organization of the genetic material of bacteria." As Klein wrote to Lederberg, "The committee had no choice but to break it up in this way."

Now that the decision was made, there was no need to make a fuss about it. As to the formalities, the Lederbergs would be invited to a dinner given by the Swedish King in his palace and meet with representatives of the Foreign Ministry at the American Embassy. Eva Klein wrote to Josh about the formalities of the occasion: he needed tails and a white waist coat, and Esther needed a full evening dress and long white gloves for the Festival and the King's dinner. She could wear the same dress, but most women wore different dresses to the two occasions (see Figure 14.1). "I am very glad that Esther can come for the sake of the whole of Sweden," Eva wrote. "To have a Prize Scientist without a lady, and especially such a lady being herself worth at least a half or more Prize would be a sad occasion. Now, this lady would have her life's occasion to make a nice fashion show for the press, but I guess that this is not Esther's ambition."[18]

Eva Klein was not alone in recognizing Esther's importance. After all, both Lederbergs had shared the Pasteur Medal in 1956 from the Illinois Society for Microbiology for "their outstanding contributions to the fields of microbiology and genetics." When Josh alone was awarded the Lilly Prize three years earlier, he told

Fig. 14.1. Esther Lederberg and Joshua Lederberg. Nobel Prize Ceremony 1958. (Courtesy of the Esther M. Zimmer Lederberg memorial Website.)

a reporter, "Esther should have been in on that too." And there were also other collaborators. "There are six to eight people in the background every time someone gets an award," Esther added.[19]

Galen Bradley, who had been a postdoctoral fellow in Lederberg's laboratory, wrote to Esther about the Nobel Prize: "Because I know the important role that you have played in Josh's personal and scientific life, I cannot consider this recognition of Josh as a sole triumph, but one of mutual effort and contribution... I recognize Josh's brilliance and his devotion to his chosen field, but I equally recognize your unquestioned talent. It is to the better good of science that you have found happiness and satisfaction in your mutual interests."[20]

George Berry was delighted by the news and sent Lederberg a copy of his Nobel nomination letter. "Bernie [Davis] wrote most of the letter for us," he said, "I wish you could have heard the countless enthusiastic comments that were made on your lecture and your contributions to informal discussions here."[21]

No one doubted that Joshua Lederberg would be awarded the Nobel Prize one day, but some were certainly surprised that it had come so soon when he was only 33 years old, and the prize was given to him for work he had started in 1945 when he was 20. The average age for a Nobel Prize in all categories since the time they were first awarded is 59, and the most frequent age bracket is 60–64.

Although there were some mixed feelings among a few of Lederberg's colleagues about his receiving the Prize at such a young age, but Bernard Davis, who had helped with the nomination letter, had no problem with it. "This recognition, of both you and the field, is long overdue," he wrote Lederberg. "This is one case where I am confident that the award of the prize to a youngish man is not likely to prevent him from continuing to be productive!"[22] *Drosophila* geneticist

H. J. (Jo) Muller, who was 56 years old when he received the Nobel Prize in 1946, felt the same: "Congratulations on a richly deserved award finally given to a biologist while he is young enough," he wrote.[23] Tatum was delighted that Lederberg was sharing the Prize with Beadle and him: "Herewith again my congratulations, and a repeated expression of extreme pleasure that you're sharing it with us!"[24]

Lederberg was a little concerned nonetheless and heard that there were complaints. One of them came from Charles Yanofsky, who had accepted the position in Stanford's biology department that Lederberg had rejected in the spring of 1957. Lederberg hoped that there would be no bad feelings about that when he came to Stanford's medical school. George Beadle explained to him that there were "very few biologists" who thought his prize "might reasonably have been deferred a bit." He had also heard the same being said when Lederberg was elected to the National Academy of Sciences, one of the highest honors in science, the year before. "I do not share this feeling in the least," he said.[25]

Lederberg thanked everyone who had played a role in his career, beginning with his teachers at Stuyvesant High School where he had been president of the Biology Club and been allowed to take elective courses in advanced subjects.[26] He also wrote to the President of Columbia University explaining how important Francis Ryan was, how Ryan had arranged for him to work with Tatum at Yale on bacterial recombination, and how Ryan had found funds for him when the Navy's V-12 program ended.[27] He sent a copy of that letter to Ryan, who replied, "I never did and do not consider myself your Teacher. I was just less able than others to keep you down."[28]

Those who had worked hard to get Lederberg to move to Stanford were overjoyed by the Nobel Prize. "We are all excited and delighted with the prize," Arthur Kornberg wrote to Lederberg. "It involves an additional trip this fall which I am afraid you will have to put up with, but this has not been an ordinary year and you have lots of time to recover in pleasant surroundings."[29] Kornberg would be awarded the Nobel Prize in chemistry the following year. Those two Prizes brought luster and fame to Stanford's new medical school.

Lederberg's award was bittersweet at the University of Wisconsin, which had never before nurtured a Nobel Prize. It was an awkward situation and he was a little embarrassed because, after all, the university had made it possible for him to get the prize in the first place and now he was leaving for Stanford. When reporters interviewed him, he emphasized that the University of Wisconsin was "one of the finest universities in the country."[30] He told them how bacterial genetics was fundamental to medicine for which the Prize was awarded. "It's like having to learn the language before learning to read Shakespeare," he said.

Lederberg also explained that his discovery of genetic recombination in bacteria could have been done decades earlier, but the field lay dormant because of an untested assumption that bacteria simply did not possess a genetic system.[31] He and sociologist of science Harriet Zuckerman later called it a "postmature discovery" in contrast to "premature discoveries" — those that came too soon, before "the time was ripe," and were overlooked for that reason.[32]

The University of Wisconsin threw a party to celebrate the Nobel Prize before Josh and Esther left for the ceremonies in Stockholm, which were held on December 10, the anniversary of Nobel's death.[33] All of his misgivings about the award would soon fade.

The Lederbergs were treated royally from the moment they arrived in Stockholm. They were met at the airport and had escorts from the Old Swedish Nobility assigned to them for the day. The whole country was involved in celebrating Nobel Week. All of the Nobelists gave short talks. Then there was a formal ceremony in the beautifully decorated town hall where the King presented the awards. This was followed by several press conferences.

Still, the Cold War put something of a chill on the ceremonies. All Nobel prize winners that year attended except for the Nobel Prize winner in literature. Soviet poet and novelist Boris Pasternak, the author of *Dr Zhivago,* which had been rejected for publication in the USSR, was not permitted to attend. His chair was held conspicuously empty, and when giving the formal toast to the laureates, a member of the Swedish National Academy of Science commented on Pasternak's absence with an allusion to Galileo who, because of his support for heliocentrism, was condemned by the Catholic Church and kept under house arrest until his death in 1642: "Knowledge should build for men, for mankind, for humanity. Such is Nobel's message. Scientists are needed for this — and poets, and men of good heart. To genius we look for leadership, albeit humbly: with a prayer that genius will be spared — yes, even spared the fate of Galileo."[34]

Lederberg gave the first formal response to the toast. Written out beforehand, it expressed a similar mood — that the only possible justification for the Prize was the praise paid to knowledge as a transcendent social value: "The illumination of human aspirations in intellect and in charity, which transcend nationality, is then the enduring warrant of Nobel's legacy," he said. "Our presence honors his hopes for the fraternity of mankind."[35]

After participating in all the pomp and celebrations, the Lederbergs travelled to Italy to rest for a few days with their old collaborator Luca Cavalli and his wife Alba at their villa near Rapallo on the Italian Riviera. Upon returning to Wisconsin, they began packing up to leave for Palo Alto. Much of Lederberg's reservations about the Prize had vanished by then, seemingly exorcized by the joyful week of festivities in Sweden.[36] His share of

Fig. 14.2. Nobel Prize Recipients 1958, The Joshua Lederberg Papers.

half the Prize was $21,000. He donated his Nobel Medal to the University of Wisconsin. He and Esther flew to California on January 2, and he took up his position at Stanford on February 1, 1959.

But he still had one more task to perform for the Nobel Prize. He needed to write a formal lecture. According to the Nobel Prize statutes, winners were permitted to defer doing so for six months after the celebration in Sweden in December. Completed in May 1959, it was a different kind of lecture from the one Beadle and Tatum had offered, which had been on the history of microbial genetics.[37]

Lederberg didn't look back. After thanking his collaborators, and especially Esther, he turned to the future as he synthesized current knowledge in molecular biology and prophesied important directions for medicine in what would later become known as "genetic engineering."[38] In the first Nobel Prize Lecture ever given on molecular biology, he explained that DNA was the basis of the gene, and RNA was "the communication channel" between DNA and the enzyme proteins. He then postulated how this new molecular genetics might someday meet medicine by allowing for the insertion of a new gene segment to replace a defective one, calling it "directed mutagenesis."[39] Short bits of DNA, he said, might be manufactured in what he called "synthetic chemistry" and then be inserted as a specific mutagen into a defective gene in order to fix it.

There were natural precedents for this technology that he and his collaborators had discovered: *Lambda* virus in *E. coli* K-12 becomes integrated into the main bacterial genome.[40] The fertility factor F can be inherited as a cytoplasmic factor and transmitted by infection, but sometimes it is also incorporated into the main bacterial genome. Bacterial viruses can also transfer host genes from cell to cell, and in order for the transduced gene to be integrated into the bacterial genome, it must find and replace its homologous gene in the recipient cell's chromosome. In genetic transduction, new genes are not simply added to the host genome; just like in sexual recombination, they must somehow replace the old.[41]

All of this hinted at how directed mutations might occur through DNA transfer. There were major technical challenges that needed to be met before one could do this: first learn how to synthesize specific bits of DNA, and then learn how to insert them into cells.[42]

What Lederberg referred to as "synthetic chemistry" in 1959 emerged as a new field of "synthetic biology" two decades later. Site-directed mutagenesis is used to study the molecular basis of Alzheimer's disease as well as the feasibility of gene therapy approaches to cystic fibrosis, sickle-cell anemia, and hemophilia. When biochemist Michael Smith was awarded the Nobel Prize in

1993 for "site specific mutagenesis," he referred to Lederberg's Nobel lecture 34 years earlier, noting that "the dilemma Lederberg posed had been resolved."[43]

In the meantime, Lederberg went straight to work. After settling at Stanford, his laboratory switched from *E. coli* to *Bacillus subtlis* to study the means by which some bacteria types could take up and incorporate DNA from their environment. This would be just one of his many post-Nobel Prize research programs.

Chapter 15

The New World

> Another attribute was his remarkable capacity for institutional innovation. He created a department of genetics in the medical school at Stanford University. Until then, genetics had been marginal — or non-existent — in medical schools. There was a widely shared assumption in the middle of the twentieth century that genetics might be intrinsically interesting but that it would never have much practical significance for medicine. How wrong that assumption was!
>
> David Hamburg[1]

The Lederbergs landed in a whole new world when they moved to California in February 1959. It was the beginning of what would eventually be known as "Silicon Valley," a community that was conceived of and parented by tech companies and Stanford. Palo Alto itself was a rather quiet community about 30 miles south of San Francisco, essentially continuous with the other towns strung down the peninsula. Josh and Esther warmly embraced the Californian lifestyle.

Less than a year after the Lederbergs arrived, they had a house built right on the campus. They got into San Francisco about once a week for dinner, shopping, or a play, and out to an ocean beach or the redwood forests as often. Berkeley was just 75 minutes by thruway and they would visit there a couple of times a month for seminars or genetics club meetings. But Stanford itself was just about the ideal university community, as they saw it. "Frankly," Josh wrote to a colleague in November 1959, "I think I'm very lucky to have not ended up at Berkeley but nearby, to have the cake and eat it too."[2]

Stanford was buzzing with intellectual energy. It seemed to be at the cutting edge of everything new and exciting. Stanford's president, Wallace Sterling, did a lot to build the university up, stripping it of its former "country club" reputation. He and provost, Fredrick Terman, erstwhile engineer and one of the so-called "Fathers of Silicon Valley," had decided that when the medical school was moved from San Francisco to the Palo Alto Campus, it would not be a typical trade school but rather comprised of scientists who would interact with colleagues in departments outside of medicine (Chapter 15). That interdisciplinary mindset fit Lederberg to a tee.

Stanford was different from Wisconsin in almost every way imaginable, not only because the long dark winters in the agricultural Midwest contrasted so sharply with the sunny California desert, but also because the culture there generally was fresh and new. California was not rooted in the past; it seemed to have the ethos of a new frontier that drew people from across the country and across the globe.

Being Jewish, for example, may have been rare at the University of Wisconsin, but not so at Stanford. Many of the leading scientists there were Jewish, as Gustav Nossal noticed when he arrived by ship from Melbourne in August 1959 on a two-year fellowship to continue work that he and Lederberg had begun on the clonal selection theory of immunity Cchapter 11). Although Nossal's father was Jewish, he had been brought up Catholic." He had been reticent about this Jewish ancestry until he arrived at Stanford where there were so many brilliant Jews. There, for the first time, he felt proud of his Jewish side.[3]

A Pocket-size Department

Unlike Arthur Kornberg, who brought his whole biochemistry department to Stanford from St Louis, Lederberg constructed his genetics department from scratch. He wanted to keep it small with a handful of faculty members and research fellows, and he aimed for a department with a diversity of intersecting research programs that centered on molecular biology and genetic aspects of immunology.

The research in his own laboratory focused on developing the foundations of genetic engineering, whose potential for medicine, in his view, was enormous: isolating desirable genes, transferring them into bacteria, and making both the desired proteins and medical products derived from them. It would take until the 1970s to achieve these goals. It was clear to Lederberg that what needed to be studied was the process by which some bacteria naturally acquire DNA from their environment — as in the famous pneumococcus transformation experiments, which had pointed to DNA as the basis of the gene in the first place (Chapter 4).

Such transformations do not occur naturally in *E. coli*, so when Lederberg arrived at Stanford, he switched to studying *Bacillus subtlis* in which they do occur. This research program began with his PhD student, A. T. Ganesan, who arrived in 1959 from unfortunate academic circumstances in India. He had been a doctoral student at the Indian Institute of Science in Bangalore, but was not permitted to complete his PhD because he had challenged some of his professor's erroneous views about chromosome behavior in yeast. Ganesan's supporters approached Lederberg, who had been to India two years earlier (Chapter 12).[4]

Lederberg readily accepted Ganesan.[5] When Ganesan completed his doctorate three years later, Lederberg regarded him "as one of the most intelligent and enterprising graduate students" he had ever had.[6] Ganesan subsequently became a member of the Stanford faculty in Lederberg's genetics department.

The Genetics of Immunity

There was still another kind of biological engineering that was developing in the 1960s. Organ and tissue transplantation could clearly be seen on the horizon and Lederberg wanted to advance it. The major barrier to organ transplantation lay in the recipient's immune system, which would treat a transplanted organ as "non-self" and immediately or chronically reject it. Some cells and tissues were also more amenable to transplantation than others. Before moving to Stanford, Lederberg searched for "bright young researchers" working on tissue genetics and immunology.

At first, he had his eye on Avrion Mitchison, who had studied with Peter Medawar, "the father of transplantation," at University College, London. Medawar shared the Nobel Prize in medicine with Macfarlane Burnet in 1960 for the discovery of acquired immune tolerance. However, Mitchison was unavailable, so Josh turned to Leonard Herzenberg, who had worked at the National Institutes of Health in Maryland. He had an excellent pedigree and had worked at the Pasteur Institute in Paris as well, but he was said to have "personality problems" that were not adequately compensated for by scientific talent.[7] Lederberg paid no heed to that; he considered Herzenberg to be "a rare bird" who shared his "own predilections for setting tissue culture on a sounder basis."[8]

Herzenberg arrived at Stanford with his wife, Lee Herzenberg, who was also a research biologist, and their two children, soon after the Lederbergs.[9] Lee worked alongside her husband as a research associate focusing on tissue cell genetics. The first successful kidney transplant occurred just as they were getting the tissue cell genetics program underway at Stanford. Using technology from an instrumentation laboratory that Lederberg had established, Leonard Herzenberg developed the cell sorter, one of the most important instruments that transformed cell tissue studies (Chapter 17).

Lederberg not only handpicked his faculty, he also played a key role in directing them down career-long paths, along which they became outstanding researchers. He had a deep scientific knowledge of just about everything it seemed. As Sir Walter Bodmer, who joined the department in 1961 as a young postdoctoral fellow put it, Lederberg was like "a maestro conductor for the whole department, who knows everything about what's going on." He had "a total recall memory, an enormous store of wisdom and interests, you could discuss anything with him."[10] He read ferociously and extraordinarily quickly. Some of his collaborators at Stanford suspected that he had a photographic memory.[11]

Bodmer came to Stanford in 1961 as a postdoctoral fellow to work on the molecular biology of *B. subtlis*. But Lederberg pointed him to a whole new field where he pioneered the genetics of immunity in humans.[12] He was trained as a population geneticist, the PhD student of well-known evolutionary geneticist Ronald Fisher at Cambridge. But he wanted to move into molecular genetics.[13] Moving to Stanford could not have been a better choice for Bodmer. He taught the first course in genetics at the medical school, and Lederberg got him an appointment as a faculty member the next year.

Bodmer worked on the molecular biology of *B. subtlis* during the first five years, sometimes collaborating with Ganesan. While Kornberg and coworkers at the neighboring biochemistry department studied how enzymes function in DNA replication outside the cell in a test tube, Bodmer focused on where and how DNA replication occurs within the cell, how foreign genes are acquired, and what the molecular basis of genetic recombination was.[14] But Lederberg had yet another idea for exploiting Bodmer's talents.

Bodmer had planned to give up mathematics and population genetics for molecular biology until Lederberg convinced him otherwise and orchestrated a collaboration that led to a whole new field, the genetics of immunity in humans. Bodmer investigated the gene complex, which encodes the white blood cell surface proteins that are responsible for regulating the immune system. It is the means by which the immune system recognizes foreign proteins (antigens), such as those on the cell surface of bacteria or of viruses.[15] It is essentially the system for recognizing self from non-self. Just as red blood cells are lined with proteins which are used to identify a person's blood type as A, B, O, or AB, the proteins of white blood cells comprise another set that must also be matched for successful tissue and organ transplantation.

Today, when a potential transplant donor is identified, the donor's white blood type is determined to assess for compatibility with the recipient's blood type. Nine key genes control these leukocyte proteins; all of them are clustered together on chromosome six in humans. None of this was known in the early 1960s. With the era of human organ transplantation just emerging at that time, the genetic study of human leukocyte antigens was critical for determining compatibility. It was also vital for our understanding of autoimmune diseases. People with certain leukocyte antigens are more likely to develop type I diabetes and rheumatoid arthritis, for example.

That genetic work began when Lederberg introduced Bodmer to hematologist Rose Payne. She was a Senior Scientist in the Department of Medicine, and she had gathered an enormous amount of raw data on white blood cell antibodies. Those antibodies are formed in patients as an immunological reaction to blood transfusions of the wrong blood group, for example. The white blood cells of those patients clump together. White blood cells also clump in pregnant women as a reaction to foreign leucocyte antigen proteins encoded by the DNA that the fetus inherits from its father.

Payne was able to accumulate an enormous amount of data from the blood of pregnant women at Stanford's Hospital in the School of Medicine, and genetic differences in father and mother could be recognized by the clumping of white blood cells. But as a hematologist, she did not really know how to conduct a genetic analysis of the data. When Lederberg saw Payne's data, he thought this was something to which Bodmer could apply his knowledge of population genetics. Payne readily agreed to collaborate, and Bodmer immediately knew what needed to be done.[16] The quantity of data meant that they needed a computer for recording and analysis.

Bodmer's spouse, Julia, who stayed home to take care of their three children, was interested in part-time work and therefore helped in establishing the new field. Although she was not trained as a geneticist, she had some statistical knowledge from her economics courses at the University of Oxford. Payne and the Bodmers reported their preliminary genetic analysis of the human leukocyte system in 1964.[17]

The Bodmers returned to England in 1970. Walter took up an appointment as Oxford University's first Chair of Genetics and he continued his work on the human leukocyte antigen system, somatic cell genetics, and human genome mapping. Julia

Fig. 15.1. Walter Bodmer and Julia Bodmer, 1970. (Courtesy of the Esther M. Zimmer Lederberg memorial Website.)

Bodmer was appointed Research Officer in the Genetics Laboratory at Oxford and continued with her research on human leukocyte antigen types, their distributions within populations, and their association with diseases such as juvenile rheumatoid arthritis.[a]

[a] Stanford was sluggish in promoting women. It may have had a modern, inclusive culture in some ways, but like many organizations and universities of the time, those attitudes were not always extended to women. Whereas Walter Bodmer had been promoted to Full Professor in 1968, Rose Payne would not be given a position as professor of medicine until 1972 at the age of 63. "I had never asked for anything," she later said. "Then, one day, I decided I was old enough and established enough. So I marched right in and asked."

Neurobiology

Lederberg had a knack for seeing how methods used in one field of research could be applied in a different one, pointing young researchers along paths, and creating fruitful collaborations as he did with the Bodmers. He had a similar effect on the career of another British scientist, Eric Shooter. Shooter had arrived at Stanford in 1961 as a research fellow studying the biochemistry of DNA replication in Kornberg's biochemistry department. Lederberg hired him as a professor in his genetics department in 1964 after he had steered him into a wholly different research direction — the brand new field of molecular medicine that Lederberg had envisaged. Eleven years later, Shooter became the founding chair of a new Department of Neurobiology at Stanford.

Shooter's aim as a research fellow in Kornberg's biochemistry department had been to provide experimental evidence that the replicating unit of DNA is the single strand. This was no great insight and rather dull in Lederberg's view. It seemed obvious from the structure of the double helix, which unwinds when it is replicated. Each strand acts as a template in the formation of a new strand.

What Lederberg did find interesting was the research that Shooter had conducted in London on the molecular genetics of hemoglobin before arriving at Stanford.[18] Molecular studies of hemoglobin began in 1949 when Linus Pauling and his collaborators at Caltech used electrophoresis — a technique for separating electrically charged molecules, such as DNA and proteins, according to size and charge — to study the structure of the abnormal hemoglobin of individuals with sickle cell anemia. It was the first so-called "molecular disease."[19] Seven years later, it was reported that the hemoglobin protein of sickle cells differs from that of normal cells in just one amino acid. A single gene mutation caused the disease.[20] Shooter and his collaborators subsequently analyzed blood samples taken from various immigrant populations in London and were able to classify some of the different forms of hemoglobin and identify the genes associated with them.[21]

When Lederberg heard Shooter give a talk on this research, he had a completely different subject in mind — the brain. As Lederberg saw it, one could do the same sort of analysis: extract proteins from the brain, separate them by electrophoresis, and identify mutations by looking for those whose charge has changed. He thought that one could study some of the obvious neurological impairments such as "mental retardation" that way. He also suspected that such studies might help in understanding how information is stored in the brain.

Shooter had to return to England when his fellowship in Kornberg's department ended, but before he left, Lederberg arranged for him to return and join his genetics department. In the meantime in 1962, Lederberg wrote a grant application to the National Institutes of Health for a program on "molecular neurobiology to search for modified proteins of the brain."[22] His grant was approved and he got a technician to begin the research. When Shooter returned to Stanford in 1964, he inherited that NIH grant and would renew it for more than 40 years.[23]

Lederberg further advised Shooter about a newly discovered protein called "nerve growth factor," which somehow promoted the growth and survival of certain neurons that transmit sensory information from the internal organs of the body back to the brain.[24] Rita Levi-Montalcini at Washington University in St Louis had discovered it in the early 1950s (she would be awarded the Nobel Prize in medicine in 1986).[25] In the early 1960s, when no one knew what the nerve growth factor protein was, Lederberg went to St

Fig. 15.2. Emmanuel Burgeon, Erik Shooter Rose Payne, 1970. (Courtesy of the Esther M. Zimmer Lederberg memorial Website.)

Louis to meet with her and learn more about how to identify it. She was eager to have that research continued in his department, so Lederberg arranged for Silvio Varon, who had worked with her for a year, to join Shooter when he returned to Stanford.[26] After three years, Shooter and Varon were able to purify the protein and determine how it affects nerve cells.[27]

Chapter 16 Molecular Medicine

> Of all the species on earth, man is in many ways the least well understood as a biological object.
>
> Joshua Lederberg, 1966[1]

Stanford's medical school was long on talent but short on space. Research conditions were cramped and people were frustrated. When Lederberg arrived in February 1959, his laboratory was in a somewhat dated biophysics building and comprised two large rooms stuffed with workbenches that were covered with agar plates, pipettes, esoteric tubing, autoclaves, handheld counters, and other paraphernalia. Esther Lederberg, and other researchers and assistants were crowded in there with him.[2] Fortunately, Lederberg had a flair for not only envisioning new fields of study but also raising funds to support them.

The Kennedys

Four months after Lederberg arrived in Stanford, he received a letter from the Joseph P. Kennedy Jr. Foundation explaining that it was interested in funding a research center focused "on those diseases which produce mental retardation."[3] The Foundation's director, Eunice Kennedy Shriver, had put a high priority on supporting research on mental deficiencies (with which her oldest sister Rosemary was afflicted).

Neither Lederberg nor anyone else had any idea how to tackle the problem directly. It was obvious to him that the affliction had roots in genetics, biochemistry, development, and the neural processes that underlie behavior. But that was about as far as he could see.[4] Little was known about "mental retardation" or human genetics at the time. It was not even known that human cells had 23 pairs of chromosomes and not 24 as had been assumed until 1956. Next to nothing was known about mental disabilities or what caused them. Several different disabilities were lumped under the rubric of "mental retardation" and, as Lederberg explained, they would need to be distinguished before one could focus on specific problems.[5]

It had just been shown that "mongolism," now known as Down Syndrome, had nothing at all to do with Mongolian peoples. Instead, individuals with the mental disability have 47 chromosomes instead of 46 due to an accident in the separation of chromosome 21 while the egg matures in the ovary. There were only two other genetically based intellectual disabilities that were well known at the time. One was phenylketonuria, which results from decreased metabolism of the amino acid, phenylalanine. The other was galactosemia, which results from an inability to digest the sugar galactose because of the presence of homozygous recessive mutant genes.

In Lederberg's view, the Kennedy Foundation's proposal to focus directly on "mental retardation" was far too narrow and restrictive. So he made a counter proposal to support a fundamental research program in "molecular medicine." He began to deal closely with The Kennedy Foundation when John F. Kennedy was elected President. Kennedy sought to forge productive relations with writers, composers, philosophers, and scientists, and he invited Lederberg to his inauguration celebrations in January 1961. Lederberg attended with the optimism shared by many that Kennedy's election promised a new, youthful, and progressive nation.[6]

Lederberg joined Kennedy's "Panel on Health and Social Security for the American People," which recommended improved medical care for the aged and children of unemployed parents, as well as improvements in medical education and research. And when Kennedy announced in October 1961 that he was appointing "a panel of outstanding scientists, doctors, and others to prescribe a plan of action in the field of mental retardation," Lederberg met with him at the White House as a member of that panel as well.

He also met with Eunice Kennedy Shriver, who had encouraged her brother to make intellectual disabilities a priority for his administration, and a few days later he had long discussions with her husband, Sargent Shriver, who headed the Kennedy Foundation. Shriver had helped to establish the Peace Corps and he was the founder of Job Corps, Head Start, and other programs of the "War on Poverty" in the 1960s. His discussion with Lederberg would have major repercussions for Stanford's medical school.

They met on a Saturday afternoon in Shriver's office at the Peace Corps. Shriver asked Lederberg about the genetic code, which was beginning to be deciphered at the time, and what all the excitement was about. As Lederberg told him the story about molecular genetics and its implications, Shriver became enthralled. He set aside appointments for three hours, asking penetrating questions about what it meant for our philosophical concept of human nature, and its potential implications for problems that the politician might have to face.[7]

Shriver came to appreciate the broad base of science that had to be developed to address mental illness, but Lederberg was still uncertain about an approach to neurobiology. Then, two weeks later, it came to him in a flash when he heard Eric Shooter give a seminar in the biochemistry department about his previous research in London, where he worked on sorting out major groups of the blood protein hemoglobin (Chapter 15). Lederberg's aim, as he wrote to Shriver on October 26, 1961, would be "to start a rather broad program in neurochemistry, to sort out the major groups of proteins found in the brain, by analogy with the analysis of blood protein that has been so fruitful."[8]

The research program itself did not need the Kennedy Foundation; Lederberg got a government grant for it. What was needed was proper space and facilities to do the work, which was also true for his colleagues in obstetrics, pediatrics, and biochemistry. As he explained to Shriver, "There are dozens of us who have ideas and programs that we are eager to pursue and are frustrated by the limitation in working space."[9]

Negotiations for funding were finalized by the end of 1961 to help build a new medical research building.[10] The Kennedy Foundation offered a grant of one million dollars, and the National Institutes of Health provided a matching grant. The entire building was expected to constitute "one of the nation's most advanced facilities for medical research." The third floor was designated as "The Kennedy Laboratories for Molecular Medicine" with Lederberg as its director.[11] Lederberg and colleagues moved into those quarters in 1966.[12]

NASA

One of the oddest things that Lederberg established in the genetics department was a NASA-based instrumentation laboratory focused on the search for extraterrestrial life. Recall that he had sparked NASA's interest in searching for extraterrestrial life when he coined the term exobiology and suggested some experimental approaches to it in 1960 (Chapter 11).[13] NASA organized a Life Science Research Office that same year and asked Lederberg to establish an instrumentation laboratory for designing the technology that could test for life on Mars in the event of a mission.[14]

Lederberg was lucky to recruit Elliot Levinthal as the laboratory's director. Levinthal had received a PhD in nuclear physics from Stanford in 1949 and had great experience in building technology. He had established an electronics firm that designed and built very specialized modulators, large transmitters, and medical instrumentation. He made a fortune, but sold that company to join Lederberg's genetics department in 1961 to set up NASA's instrumentation research facilities.

The NASA lab had a staff of eight people whom Lederberg worked closely with to plan the technology. "There can't be a meeting here," Levinthal said, "without Josh raising the most important and provocative issues."[15] Their aim was for an automated computer-based system installed on a lander to detect life on Mars. Mass spectrometry seemed to be the best chemical device for the job. It would separate components from the environment to indicate their chemical composition.[16]

Lederberg designed what he called "the Multivator," and Levinthal and his assistants made a prototype. A conveyer belt would scoop up samples of Martian soil and deposit them into a computer-controlled mass spectrometer. The soil was bombarded with electrons inside the spectrometer to produce a fragmentation pattern,

Fig. 16.1 Stanford Genetics Department 1963.

which sorted the electrified particles (ions) according to their mass. This pattern was to be transmitted to earth where scientists could analyze it for evidence of organic compounds and microbial life.

The development of the mass spectrometers and computer systems attracted a flow of visitors to Stanford from 20–30 institutions a year by 1970. A commercial version of their computer control interface was being produced at that time by Systems Industries in Sunnyvale, California.[17] Levinthal and his team tested for carbon compounds in two samples from Apollo 11, the first human landing on the moon in July 1969, for carbon and associated organic compounds which, Lederberg had proposed, might have been brought there in meteoritic dust. But all they detected were carbon compounds that were probably brought there by the landing of the lunar module itself.[18]

The instrumentation research laboratory also contributed to the design of experiments when the Viking Landers went to Mars in the summer of 1975, taking photographs and collecting other data on the Martian surface. The two landers conducted three biology experiments designed to detect signs of life in Martian soil based on signs of enzymatic activity. They detected enigmatic chemical activity but provided no clear evidence for the presence of living microorganisms in soil near the landing sites.

It might seem inappropriate to have NASA-funded research in a medical faculty, but those aims dovetailed in a strikingly innovative way. In fact, the laboratory had been charged with biomedical application from the very beginning.[19] Lederberg was able to write to grant agencies in 1968 that the program in automated instrumentation was "converging with the main scientific aims of the department."[20]

As it turned out, NASA's lab had a great impact on medicine with the development of a technology for diagnosing, monitoring, and treating cancer and infectious diseases. It is called the fluorescent activated cell sorter, an instrument for sorting out and isolating single cells to be studied independently. When a patient has cancer, for example, the sorter can separate the rare immune stem cells that give rise to many other kinds of blood cells from a patient's blood. Many experiments in modern immunology are also possible only because of that cell sorter. It is one of the most important devices ever developed for clinical research in genetics and immunology.[21]

Separating and sorting cells isolated from human blood or tissues according to their type had been an extraordinarily tedious and difficult task. Cells were typically separated based on differences in cell size, shape, and surface proteins. The cell sorter developed by the NASA lab changed all of this. It could analyze, and sort cells from small samples of human blood cells, for example — just like a coin sorter can separate a bag of change into neat stacks of quarters, nickels, dimes, and pennies. But rather than separating cells by size, the cell sorter divvied them up according to fluorescent labels attached to the cells' surface: green cells shunted into one tube, red into another, and unstained cells into yet another. Those labels are matched with specific antibodies that target and attach to proteins found only in certain cell types.

Levinthal and his NASA instrumentation group built the cell sorter in collaboration with Leonard Herzenberg. Lederberg had hired him for his research on tissue culture because of its medical importance for immunology (Chapter 16). But the biological problem that set Herzenberg on his path toward the cell sorter was more personal: his eyes hurt. He later recalled, "I was sitting in the lab one day counting immunofluorescent cells under the microscope, and I said, 'There's got to be some kind of machine that can do this.'"[22]

The origin of the cell sorter actually began at the site of the Manhattan Project in the Los Alamos National Laboratory, New Mexico where Physicist Mack Fulwyler invented a machine to sort cell-sized particles by volume. Initially, he was working on a project to monitor fallout in food from atmospheric nuclear weapons testing.

But when the treaty banning nuclear weapons testing (except those conducted underground) was instituted in 1963, he turned to applying the technology to biomedicine. Two years later, he reported that he had improved it and used it to separate blood cells from mixtures of human and mouse blood.[23] Herzenberg headed to New Mexico to see him and brought blueprints of the machine back to Stanford, where he collaborated with Levinthal and his engineers to build the fluorescent activated cell sorter.[24] They published their first successful results in *Science* in 1969.[25]

As Lederberg saw it in 1971, the cell sorter's immediate research value was for investigating the immune system and understanding tissue rejection from transplants, the very target he was aiming for when he had hired Herzenberg in the first place.[26] That was an understatement. Herzenberg collaborated with a company to commercialize production of the cell sorter and established a core facility at Stanford to ensure that everyone had access to it. He was awarded the Kyoto Prize, which is regarded as Japan's equivalent to the Nobel Prize, in 2006.

Hence, having NASA's instrumentation laboratory to detect life on other planets in a medical faculty, as odd as it might have seemed to outsiders at the time, paid off hugely. But its benefits to the medical faculty did not end with the development of the cell sorter. It led to the development of a computer system in Stanford's medical faculty that was far more advanced than those in any other medical school. It also had major repercussions for Lederberg himself. Developing the automated technology to identify organic compounds found in space led him into another parallel career in artificial intelligence (Chapter 18).

Chapter 17

Breakup

> So in the past decade I have done my job of being a successful scientist and perhaps not so successful husband.
>
> Joshua Lederberg, diary entry, January 3, 1959[1]

Lederberg's first five years at Stanford were intense as the young Nobel laureate was, as he put it, "transformed from a private person to a semi-public institution."[2] He was sought after for guest lectures at universities and international conferences, and he was on hiring committees for the new medical faculty, building up his small genetics department, and starting a new program of molecular genetics in his laboratory. He advised his young faculty on research directions, helped establish collaborations for them, and worked closely with the NASA laboratory to design instruments for detecting life on Mars. He suggested to engineers at NASA's Jet Propulsion Laboratory that rocket ships could be fueled by "crystalline plasma."[3] He also advised on boards of tech companies and he fired off ideas on computer applications for science.[4]

National research panels also looked to him for advice. But his heaviest extracurricular commitment was with the National Academy of Sciences' Space Science Board that advised NASA. He just seemed to be in his element; he had found a place where his transdisciplinary intellectual talents could be given free rein.

Despite all of this, Lederberg's self-confidence dramatically declined during the five years following his Nobel Prize. Self-doubt threatened to overwhelm him. He fell into a depression rather suddenly, just before moving to the West Coast.[5] Much of his anxiety stemmed from his troubled marriage. He had kept a diary when he was a lonely schoolboy. He read Freud at the age of 12 and delved into introspection for a while. He dropped the diary during his undergraduate days, but partially took it up again in January 1959, when he began to write about his relationship with Esther.[6]

At first, he said, their marriage nurtured real warmth and sensitivity, but the relationship started to crack after only a few years before it broke altogether. As he saw it, there were multiple causes: friction from intersecting careers, her over-dependence,

Fig. 17.1. Esther Lederberg and Josh Lederberg, circa 1962. (Courtesy of the Esther M. Zimmer Lederberg memorial Website.)

his lack of affection, and her persistent derision and ridiculing of him especially when his professional interests broadened far beyond the laboratory. In addition, the couple was unable to have children.

The years between 1947 and 1950 were the happiest period of their marriage as they settled into a bucolic life in Madison. They were eventually able to get a mortgage on a house as his starting salary of $3,600 increased rapidly. But domestic harmony was broken when Esther was completing her thesis work in 1950, mainly, he believed, because of his involvement in it. He had encouraged her to do the PhD, and though he was not her official supervisor, she worked on bacterial genetics under his supervision (Chapter 8).

Esther asked him to help with both the writing and the defense of her thesis. "Esther," he said, "was a careful worker and perceptive observer." But while her keen observation and technical skills were unrivalled, her critical thinking skills weren't as sharp. "She was less capable — at the very least totally lacking in self assurance — at critical hypothesis-formation and testing. The experiments she pursued for her thesis were gratifying; but writing her dissertation and defending it were an agony, I had to take unwanted part in it, and how to evade or acknowledge this became a major stress between us."[7]

Further strain on their marriage occurred when they tried to start a family along the same happy pattern they had seen among their friends in Madison and elsewhere. Many wives continued to work as outstanding scientists after having children. Harriett Ephrussi-Taylor was a case in point. Josh had known her since medical school days in New York and had been infatuated with her, but she was seven years older and showed no interest in him. She married geneticist Boris Ephrussi, a man 17 years older than herself, and moved to Paris where she continued her laboratory research and raised a family.

Josh Lederberg's good friend Evelyn Witkin, a leading bacterial geneticist at Carnegie Institution's Cold Spring Harbor Laboratory, had a child a year after she completed her PhD in 1948. "Being a mama is lots of fun," she wrote to him in August 1949, "our little Joey grows and grows, and brings us lots of joy of a brand new type 1. I must confess I had some anxiety at first about being a part time mother, but so far at least it has been working out wonderfully well. No signs of neglect yet either in Joey or the little bacteria."[8] Witkin became one of the world's leading molecular geneticists.

The Lederbergs foresaw their own future in a similar way. They planned to start a family after Esther earned her PhD. But, for Joshua at least, it secretly promised to "separate the conflict of our intersecting careers, and enhance Esther's sense of purpose in life."[9] They had stopped contraception in the summer of 1950 *en route* to California where he was to give a series of lectures at Berkeley. Sadly, month after month, they had no luck.

When they returned to Madison at the end of that summer, Esther took up a position as Research Associate in Josh's laboratory, and, in his view, that only made matters worse. When students and fellows arrived, she was "*de facto* executive officer" for the group. "The boss's wife" is always in a different position and, as he saw it, the "combination of Esther's ambition, insecurity and jealousy for my time made it more so." He had hinted that she would be better served working in a more independent role, but as he recorded it decades later in autobiographical notes, "She took such hints very badly," and he never enforced them: "My never articulating what I wanted from her was ultimately demeaning: the answer became, to be let alone i.e. nothing from her was of real value."[10]

Perhaps. But Josh's negotiations with Berkeley in the late 1950s seem to belie his recollection of a one-sided dependency. Their collaborative research had almost come to an end by then, but he was rather adamant that she work in his laboratory, arguing that she was critical to his maintaining broader intellectual interests outside the laboratory (Chapter 13). She effectively ran the laboratory, arranged things for visiting researchers, and oriented postgraduate students and postdoctoral fellows. Whatever the case was in the laboratory, he was resigned to a marriage "that oscillated between neutrality and hostility."[11]

Divorce was out of the question in the 1950s. It "seemed unthinkable, legally impossible in any dignified way and particularly were she to resist as she would." Nor, in truth, could he conceive of any alternative but to stay in the marriage. If she was emotionally overly dependent, the same was certainly true of him. He had been lonely for so much of his life that he did not want to be alone again. He had also gained a great deal of weight in the 1950s. "With my corpulence could I attract anyone 'better' than I had? Esther had fulfilled my deepest expectations about women and about myself — and therewith some neurotic needs about what human life was about."[12]

Josh had difficulty interacting lightheartedly. He was not good at chitchat. His friends, James and Ann Crow, seemed to be paragons of marital happiness, but he could not be content with "the level of intellectual discourse (or underplay) in which they could engage."[13] His marriage suffered from the very thing that made him appear so impressive to his colleagues and students. Gaylen Bradley, one of his postdocs at Wisconsin, recalled that when the Lederbergs hosted weekly lab meetings in their home to discuss recent advances in microbial genetics, Josh would sit silently on the floor listening before creating an impressive finale by making some sort of sense over his students' interpretations of the data.[14]

Gus Nossal noticed Josh's anxiety when they worked together at Macfarlane Burnet's laboratory in Melbourne in 1957 (Chapter 11). Lederberg was at ease sitting on the floor with him in front of a radiator in a cold Melbourne apartment during the Australian winter talking about science and life. Nossal was a first year graduate student then while Lederberg was ten years a scientist, founder of a new discipline of bacterial genetics, and destined to win the Nobel Prize. Yet, he treated him as an equal colleague in every way. "And I found it interesting," Nossal recalled, "because he never, he never took advantage of the asymmetry of the situation."[15] Unlike Mac Burnet who was a slow methodical thinker, he said, Lederberg "was lightning fast" and he "got a lot of his sparks through the thrust and parry of vigorous discussion. He loved it, and he was so good at it."[16]

Still, Nossal felt that something was eating at Lederberg even then. He was nervous. "He had very badly bitten nails, he must have had somewhere, a nervousness inside him, a nervous tension of some sort inside him, the smartest person I ever met." Something was not right. Nossal recalled that "Esther was lovely, but [...] she really had logorrhea. She was a compulsive talker. She would talk and talk and talk and talk and talk till she'd drive everyone crazy. And she must in the end have driven Josh crazy. But she was a lovely person and a motherly kind of person. She never had any children. And I think that might have had a bit to do with the breakup of their marriage because he wanted children."[17]

By all accounts, Esther was jovial and personable outside the laboratory. She was also extremely helpful in organizing functions for students and making arrangements when new faculty arrived with families. But inside the Laboratory, she was very strict about the way things ought to be done; she was someone to stay clear of as much as possible.[18]

Psychoanalyst Clay Whitehead recalled his experience as a teenager working for two summers in their Stanford laboratory soon after they arrived.[19] Esther defined his tasks and showed him the proper way to wash test tubes and prepare agar in petri dishes. "She seemed to exude qualities of kindness and competence combined with a sort of dour dysphoria which caused this insecure adolescent to keep a distance." Josh, however, was different: "You would not have guessed he was a preeminent scientist from his demeanor, which was unpretentious and slightly detached... When not interpersonally engaged, he seemed perhaps distracted; but when engaged, he was resonant, generous, and available intellectually. He radiated energy, was voraciously curious, and struck me as driven and hyperkinetic."[20]

Nonetheless, Josh saw his inability to engage lightheartedly with others as a deep flaw in his character. "Even now," he wrote in his diary notes in 1959, "I find myself using mainly objective criteria in interpersonal relations. So a cold fish?... As a youth I was painfully lonely, as I could easily be now except for my cherished Esther, and much of what came about must have been a reaction to this. And I suppose one can trace all of this dissociation to my intellectual precocity... But lonely, lonely it was, and plain even now to see or read."[21]

Esther did not fail to remind him about his lack of basic interpersonal skills. As a result, he had fallen "under a malignant spell where Esther's disparagement of me became all too believable. At least in interpersonal relations, and my passive acceptance of that seemed to be sufficient proof of my own deficiencies. So," he said, "I continued to minimize that sphere and seek rewards in my work."[22] Minimizing interpersonal relations for intellectual work no doubt is a trait of many a scholar. In Lederberg's case, it was perhaps chronic.

The Lederbergs' marriage weakened as his intellectual interests broadened beyond the laboratory, especially when he began to participate in national and international policy issues while she remained dedicated to laboratory life. A telling episode occurred on June 22, 1958, several months before news of the Nobel Prize and before the move to Stanford, which Josh memorialized in a diary entry.[23] He had received a phone call followed by a formal letter inviting him to participate "in a top secret study and work group" for the Advanced Research Projects Agency (ARPA) of the Department of Defense.[24] ARPA was assigned responsibility for space technology, defense against ballistic missiles, and any other advanced research that might be designated by the Secretary of Defense.

He was asked to join a group of scientists to work on "the identification of advanced projects" vital to national security which, he assumed, meant "speculative weapons systems and counter-measures."[25] Scientific involvement in such matters had been typically reserved for physicists and chemists. It was rare in those days for a biologist to be asked to participate. The briefings would be top secret.[a]

Lederberg weighed the pros and cons of accepting. On the pro side, there was his personal debt to national defense — his protected status as a medical student in the Navy during the Second World War.[26] On the con side, the whole thing "might well be nonsense and window dressing," and he would also have to go

---

[a] Their intention was to establish a national science laboratory with the aim of helping the United States compete with the Soviet Union in the intermediate range ballistic missile race.

through security clearance. But that was not a big issue and he had done that the year before. He was also a bit concerned that having access to classified material might restrict his freedom to publish, criticize, and make open comments on space programs.

But all of this paled in comparison to the reaction from Esther and the potential "domestic crisis." She told him that she would "take strenuous measures" to prevent him from taking on any such commitments in security matters. She ridiculed his sense of patriotic obligation and derided him as an "ineffective and unorganized person" who would have been unlikely to contribute anyway. What could he do about thermonuclear war? If he did this work, he could be "doing a sloppy job both for the government and in the lab." In sum, she saw this as "a futile enterprise, fraught with trouble for my [his] career work and for our domestic relationship."[27]

In the end, Lederberg chose not to accept the offer to work for the Defense Department. He simply could not manage it right then. He was finalizing the negotiations at Stanford, and it was important to see if the domestic crisis could be tempered and if he could make reasonable compromises by dropping other commitments.[28] He knew though that he was simply deferring the problem, which he suspected would "recur many times in different forms." As indeed it did.

Beginning in the 1970s, he did work as a consultant to the Arms Control and Disarmament Agency of the State Department on the control of chemical and biological weapons and was active on the Defense Science Board (Chapters 27 and 28). The decision not to join ARPA in the summer of 1958, he wrote to a colleague, was "colored in part by (then) marital circumstances, of such is War and Peace."[29] He also became a member of the Jasons, an elite group of thirty or so scientists who advised the government on matters of science and technology, mostly of a sensitive nature.[30]

If Lederberg's activity in international affairs was a problem, the Nobel Prize in the fall of 1958 only added fuel to domestic strife. He had always cultivated a kind of "asceticism and introversion from the human world."[31] But now that was gone, and there was an impassable barrier to getting it back. After arriving at Stanford, he was spending most of his time outside the lab. His self-confidence was ebbing and not helped by "domestic schi... or by the external 'recognitions' of my present job or the Nobel prize." If anything, he said, "I am more pressed than ever to expose myself (talk, paper, etc.) and have less and less to say."

The days when he and his little group of collaborators had the field to themselves already seemed in the distant past as bacterial genetics opened up the development of molecular biology, which in turn required wholly new technical skills.[32] So by the early 1960s, he was having a difficult time getting a strong research program going in a rapidly crowded field. And he abhorred a crowd. He considered leaving the laboratory for the library. After all, he enjoyed reading, writing, and learning as much as experimentation.

He remained in a depressed state as the world seemed to deteriorate around him. Globally, there was the ominous threat of nuclear war, "the concept of the end of the contemporary culture, if not of the species, through nuclear annihilation."[33] Domestically, conflict was fierce with Esther, and he felt like running, escaping to another country. "For the first time in years, I seriously wondered yesterday about casting off my present existence," he wrote in August 1960. "Esther and I had a furious quarrel, not speaking all day, over essentially nothing. But she is my main link to 'matter.'"[34] He had invested almost all their savings in securities ($60,000) as well as the money from the Nobel Prize with a few companies, including the pharmaceutical company Syntex, which had patented the birth control pill. So, he wrote cynically, "I will continue enjoying my do-nothing administratorship; our beautiful home, my high salary and income tax, and stock options and space boards."[35]

Lederberg had become ever more worried about his corpulence. He was no more than 5ft 11" and he weighed 253 lbs by 1960.[36] So he did something about it. By the end of 1962, he had

Fig. 17.2. Joshua and Esther Lederberg April 1958, Madision. (Courtesy of the Esther M. Zimmer Lederberg memorial Website.)

Fig. 17.3. Joshua Lederberg, in front of exobiology equipment, 1965. The Joshua Lederberg Papers.

undergone a remarkable transformation and managed to lose almost 100 pounds. He was literally a new thin man (Fig. 17.3).[37] Despite his new look, he remained in a very dark place after five years at Stanford. His creativity in laboratory research was fading as was his marriage.[38] He went out to buy a convertible (Comet '63) "for consolation."

His domestic despair was finally resolved in the summer of 1966. After twenty years of marriage, he and Esther separated and were waiting for the courts. "Esther hasn't taken this lightly (nor have I though spread over much longer time); but we are both coming along ok," he wrote to his friend and mentor Tracy Sonneborn at Indiana University.[39]

It has been suggested that the Lederbergs' marriage came to an end after he was awarded the Nobel Prize and when he headed the department of genetics at Stanford.[40] But that account is faulty because their marriage had been strained since 1950. There is little doubt that Josh's fame had been difficult for Esther. Their good friend James Crow put it this way decades later: "She was quite ambitious in her own right, and I suppose it must have been a little bit frustrating to have this genius husband… She was a good scientist on her own… Not in his class. But then he was unique, so it's no discredit to her to say that."[41]

Some commentators on Esther's career have likewise stressed that her contributions were underappreciated and overshadowed by her husband's renown.[42] If so, that was not his doing, not as he or how Norton Zinder saw it (Chapter 8). They had deliberately bolstered Esther's recognition perhaps beyond what they thought she actually deserved.[43] As one of their close colleagues stated, "They tried to make her into more than what she was as a scientist... and with the rise of feminism, there was a rewriting of history to make her even more important."[44]

This is not to say that women scientists have been given their full due in history or that they have not been denigrated.[45] Nor is it to deny that sexism existed in universities. Stanford, liberal in so many ways intellectually, had been rather notorious in its treatment of women scientists (Chapter 16). After their divorce, Esther moved into the laboratory of their old collaborator, Bruce Stocker from Wisconsin days, who had taken up a position in the department of medical microbiology in Stanford that year. She would later become the founding director of the Plasmid Reference Center at Stanford, which collected and stored a diversity of bacterial plasmids. She remained in that position until her retirement in 1985.

In 1966, the year that Josh and Esther separated, he met Marguerite Stein Kirsch, then a fellow in the paediatrics department at Stanford. She was married then with a one-year-old son, but she divorced her husband and married Josh on April 5, 1968, the day after Martin Luther King Jr. was assassinated. Marguerite's son, David, was warmly embraced by his stepfather who was enchanted by children.[46] Josh finally got the family he had always wanted.

By that time, Josh had fully reinvented himself, reemerging like a phoenix from the ashes with a new optimism and confidence about how he could further contribute to science. He had an active group of molecular biologists working in his laboratory, but his main interest by then was in artificial intelligence, "trying to teach a computer to think," and applying it to chemistry and medicine. He had also begun a weekly column in *The Washington Post* expounding on social and ethical issues arising from advances in science and technology. He was back in his element.

Chapter 18

Teaching a Computer to Think

> I have been (together with Ed Feigenbaum of the Computer Science dept.) trying to teach a computer to think about organic chemistry. (My own first pass at genetics couldn't find a good point of departure)... But we do have some very clever programs by now. One has to put as much effort into teaching them as any obstreperous undergraduates.
>
> Joshua Lederberg to Jo Muller, November 2, 1966[1]

In the mid-1960s, Lederberg expanded his research interests far outside genetics. He introduced a time-share computer system for biomedical research — unheard of at the time in any medical faculty — and he developed an artificial intelligence system, one of the most successful ever applied to science. All of this occurred in a crucible unique to Stanford.

Back in 1941, Lederberg had seen a computer punch card system used in a lab at Stuyvesant high school. He read up about computers again just after the Second World War and took a programming course in 1953 while at the University of Wisconsin. He was on the lookout for computer applications thereafter. He was primed, for example, to see the analogy of computing code and the digitized genetic code in DNA based on four letters, A, G, C, and T, grouped into millions of three letter words (e.g., AUG, GGC, ACT, etc).[2] As such, it came rather naturally for him to suggest in 1960 that the ability to store computer information in DNA would be a great contribution both to theoretical biology and engineering.[3] What he had in mind was "the molecular computer," an enormous field and opportunity that remained unexploited, he said, for lack of "a tradition binding electrical engineers with organic chemists."[4] It would be several decades before this project began to be realized.

Computing in the 1960s was mainly used for complex mathematical calculations, but Lederberg saw that it also had applications for the management of scientific information in the post Sputnik era when science was growing at an exponential rate as federal funding increased dramatically. He served on the President's Scientific Advisory Panel which was focused on that very thing in 1961.[5] Furthermore, he was by then already in the midst of helping entrepreneur, Eugene Garfield, establish one of the earliest uses of computers for data mining — a system that allows scientists to track the development of research in their fields. It was simple and ingenious. When scientists cite papers, this leaves a genealogical trail of sorts through which one can map the development of a research topic. Garfield launched what he called "The Science Citation Index" in 1964.

Lederberg played a key role in getting that project off the ground when literally no one else was interested.[6] "Since you first published your scheme for a 'citation index' in Science about 4 years ago I have been thinking very seriously about it, and must admit I am completely sold," he wrote to Garfield in 1959. "In the nature of my work I have to spend a fair amount of effort in reading the literature… and it is infuriating how often I have been stumped in trying to update a topic, where your scheme would have been just the solution!"[7] Garfield promptly replied: "I hope you won't be embarrassed by a show of emotion, but your memo almost brought tears to my eyes. It then seemed that over six years of trying to sell the idea of citation indexes had not been completely in vain."[8] Lederberg subsequently helped Garfield acquire funding from the NIH to conduct a trial run in the field of genetics.

The Science Citation Index not only functioned as an aid for scientists to stay up to date, it could also be used to assess the impact of papers and journals based on how many times a research paper is a cited and, as Lederberg suggested, would be useful to funding agencies in assessing the impact of their funding.[9] While Lederberg was good with ideas, he was a lousy businessman.[10] He had advised Garfield not to commercialize the citation index because he would lose his shirt. Garfield did allow him to make a small investment in the company though, and it was the best investment he ever made. Garfield made a fortune.[11]

Lederberg was drawn deep into the field of artificial intelligence in 1961 when he was recruited to the National Academy of Sciences' Space Science Board and into the Institute of Radio Engineers. The Institute's membership included a subscription to its journal. Its first issue, published ten years earlier, had articles on artificial intelligence by two of the field's pioneers, Marvin Minsky and John McCarthy, who had cofounded the Computer Science Artificial Intelligence Laboratory at MIT in 1958. Minsky's 1951 paper, "Steps Toward Artificial Intelligence," and McCarthy's intellect and excitement, Lederberg later said, gave him "a sense of tangibility of the possibilities of engaging in AI Research."[12]

Lederberg had met Minsky in the mid-1950s when he visited him in Madison to discuss automata and controls for a magnetic or thermal micromanipulator for the confocal scanning microscope he was constructing.[13] He met McCarthy in 1962 soon after he was hired to start up the computer science department at Stanford.[14] Stanford organized elementary programming courses to broaden interest in computers that year. Lederberg enrolled and quickly "succumbed to the hacker syndrome."

Making mistake after mistake, he vowed to "master the ****ing system!" He did mostly, he said, "out of determination not to be made a fool."[15] He later joined a social group of computer enthusiasts and met McCarthy around the computer room that summer.[16]

That's when computer-related things really got going in the medical faculty. Lederberg wanted to hire Minsky and create a "national center in the field of artificial intelligence."[17] Minsky decided to remain at MIT, but the discussions around him prompted Lederberg to bring interactive computing time-sharing to Stanford's medical school along the lines of what Minksy had helped to design at MIT. With the time-sharing model, terminals are linked by means of cables to a central computer with a multiprogramming operating system that allowed for multiple users simultaneously.

Lederberg had that kind of system built in the medical faculty at Stanford in April 1965 with a five-year NIH grant at a cost of $522,000 per year. It was named the "Advanced Computer for Medical Research" and it provided computing to researchers in the medical school.[18] It was the first community-access time-shared system at any medical school and served as the model for computing in universities during the ensuing decades.[19]

In the meantime, Lederberg had turned to developing an artificial intelligence system with the intention of applying it to scientific research. He had enrolled in courses in logic and the scientific method when he was an undergraduate student, read treatises on mathematics and on philosophical biology, and envisioned computer systems that could not only emulate the inductive reasoning of scientists (that is, drawing general conclusions from specific evidence) but also learn from experience.[20]

He aimed for this objective right after he mastered computer programming. At first, he wanted to apply it to genetics but could not think of appropriate genetic problems. Soon enough, however, he settled on using AI to infer molecular structure from mass spectrometry data.

Mass spectrometry was a technology for determining the molecular composition and structure of unknown substances. Lederberg hoped to use it in a fully automated probe to detect trace elements of life, such as amino acids on Mars, when that opportunity arose. Microgram samples would be scooped up, put into a mass spectrometer, and then analyzed by a computer program on the lander. The data would then direct the next experiment. This was Lederberg's dream.[21]

First, he had to work out an algorithm to represent three-dimensional molecular structures in the form of graphs that computers could understand, and then dissect those graphs into subgraph fragments, just as the mass spectrometer does for molecular structures. That was his task for 1963–64.[22] He was successful, and his system could create a tree chart containing all possible configurations of atoms. He called it a "dendritic algorithm" and it was one of the most sophisticated of its time.[23] Lederberg shortened the name to DENDRAL (from "dendron," the Greek word for tree, and AL for algorithm).[24] It was both a landmark program for graph manipulation by computer and the basis of the first successful form of artificial intelligence applied to science.[25]

Representing branching molecules in a graph was difficult enough, but cyclical compounds were even more difficult. "When it came to cyclic graphs," he recalled, "I had a particularly entertaining time, almost at the level of recreational mathematics."[26] In fact, it was more than idle entertainment. In doing this, he also laid the foundation for new developments in graph theory.[27]

Although the DENDRAL project aimed at developing a fully automated system to look for evidence of organic compounds on Mars, it did not turn out quite the way Lederberg had hoped. He was advisor to the Viking Landing missions of 1975 and a mass spectrometer was indeed sent to Mars, but the data were taken back to Earth to be analyzed.[28] Enigmatic chemical activity was found in Martian soil, but there was no evidence of life near the landing sites.[29]

DENDRAL was not solely aimed at outer space though. An AI system for inferring molecular structures could potentially help chemists on Earth. Before the toxicological and pharmacological properties of a compound can be assessed, its molecular structure — the configuration of its atoms — must be determined. Inferring molecular structures from a mass spectrometer was a difficult and tedious task for even the most well-trained chemist.

Lederberg wanted his AI program to be able to essentially replace a human spectrometrist. He got halfway there in 1964. While the mass spectrometer could provide a list of all the possible structures of a molecule, it could only determine the most likely structure if the unknown substance was simple. Still, Lederberg promised that when DENDRAL was fully operational, it would perform this task with greater speed than an expert spectrometrist and with comparable accuracy. To do that, it needed to be developed further by computer experts, and it would need experts in chemistry to teach the machine what they know.

By this time, Lederberg had entered into a fruitful collaboration with Edward Feigenbaum, a former student of one of AI's founding fathers, the Nobel prize-winning economist and cognitive psychologist, Herbert Simon, at the Carnegie Institute of Technology in Pittsburg. Feigenbaum's PhD research had focused on models of memory and learning, and he had created a computer program designed to simulate verbal learning.[30] He and Lederberg first met in 1963 at a meeting on Artificial

Intelligence and Machine Learning at the Centre for Advanced Study in Behavioral Science at Stanford. Feigenbaum had just published a landmark book on artificial intelligence, an anthology called *Computers and Thought*, containing twenty classic papers by those who had defined the field of AI.[31]

There were very few computer science departments at that time, and Feigenbaum was working in the Business School at Berkeley.[32] But when Stanford formed a computer science department, McCarthy recommended him for a position. When Feigenbaum arrived in January 1965, he immediately went to see Lederberg and they began to collaborate on inferring organic chemical structures from mass spectral data.[33]

Lederberg's DENDRAL algorithm gave a list of possible structures from the mass spectrometry data. But it could not narrow the possible forms from the likely ones, and for large complex organic molecules, the number of possible forms of a given atomic composition could run into the thousands. But chemists knew that certain combinations were more likely than others based on chemical theory and experience. They used mental shortcuts, or what is called a heuristic approach, to determine the structure. They used their intuition, or tacit, unspoken knowledge based on rules of good judgment, to generate hypotheses about the identity of the compound, and then designed key experiments or measurements aimed at eliminating all but one of the alternative structures.

So Lederberg and Feigenbaum's next phase of the project, which they called "Heuristic DENDRAL," was to teach the computer how to process information like chemists did. Through a dialogue between chemist and machine, the computer would learn the rules and incorporate them to produce the same hypothesis as a chemist might about what the correct structure could be. In April 1965, Feigenbaum floated a message asking faculty to contact him or Lederberg if they were interested to collaborate on the "Stanford Artificial intelligence Project."[34] But he got no replies.

Lederberg took a much more direct route and leaned on his friend Carl Djerassi to collaborate. Djerassi was one of the most famous chemists in the world, having sparked a sexual and social revolution by synthesizing the first oral contraceptives at Syntex Corporation. He also held the patent for the world's first antihistamines, and he was an innovator in the field of non-toxic pesticides. He arrived at the Chemistry Department in September 1959, eight months after Lederberg, but he also continued his industrial career as vice-president of research at Syntex Corporation and recruited Lederberg as an advisor.[35]

Djerassi was a mass spectrometrist *par excellence*; his knowledge of mass spectrometry was unsurpassable. Feigenbaum recalled the time when Djerassi first experienced their heuristic DENDRAL program:

> So we showed Carl our program and he said, "Well yeah, okay, but can it do..." and then he picked some very simple molecule, "Alcohol? Can it do alcohols?" And we said, "Carl why don't you try it?" Of course it didn't know anything about alcohols so it was really stupid. And Carl said, "but that's stupid," and we said, "Yeah that's right Carl, now what is it that you know that this program doesn't know?" And then as soon as he told us, we're typing this in; we had a presentation for it.[36]

Over the next few years, the AI system that Lederberg, Feigenbaum, Djerassi, and their collaborators put together learned to determine structures at the level of a postdoc, or even better. By 1978, DENDRAL was being used increasingly by chemists to check their structural proofs.[37]

The success with DENDRAL stimulated the development of similar programs during the 1970s. Working with medical specialists, programs were developed to aid in the diagnosis of pulmonary function and infectious diseases, and for optimizing the prescription of antibiotics.[38] It was not only used for medical diagnostics, but Lederberg and Feigenbaum also developed a successful AI program that could aid theory formation in molecular genetics.[39] "I believe it was about one year ago that I found how to reformulate the theory-formation problem," Lederberg wrote in a memo in January 1976.[40] This program would also be applied to planning recombinant DNA experiments and the insertion of specific genes into bacteria to be cloned for making protein-based drugs.[41] AI was also going to be needed for the tsunami of data to come from molecular biology. "Biology," Lederberg said in 1988, "will soon suffocate in the sheer bulk of knowledge about DNA and protein structures unless new epistemological machinery can be invented."[42]

Fig. 18.1. The Dendral team, 1991 Bruce Buchanan, Georgia Sutherand, Edward Feigenbaum, Joshua Lederberg and Dennis Smith.

Knowledge is Power

The artificial intelligence system that Lederberg, Feigenbaum, and collaborators created was an "expert system" — one that emulates the decision-making ability of a human expert based on "if-then rules" rather than through procedural computer code. The key to their success with heuristic Dendral was in endowing the computer program with *knowledge* in the form of rules and procedures to guide problem solving processes. It was not based on programming the computer with formal reasoning methods, the approach that others in AI had tended to use.

As Lederberg stated: "Knowledge, not tricks or metaphysical insight, made the program effective."[43] The idea that "knowledge *is* power" was a paradigm shift in AI and became the slogan of a knowledge-based-system movement within AI research for the next two decades. Lederberg not only helped to articulate and develop "the knowledge as power" movement in AI, he also provided conditions that fostered the growth of a new national wife computing facility for medical research.[44] Established in January 1974, funded by the NIH, and directed by Lederberg, the "Medical Experimental Computer" was eventually attached to a communication network of the Department of Defence called the ARPANET and was the precursor to the internet.[45] Stanford's medical computer lasted for about 19 years and supported more than 20 AI-in-medicine projects conducted nationwide.[46]

Email was invented on the ARPANET around 1972, and Lederberg and colleagues ran their projects without the need for face-to-face interaction 20 to 25 years before email came into general use.[47] They also posted drafts of papers and research grant proposals on common bulletin boards for the group to review and make changes to, going through scores of drafts this way.[48]

All of this would become the basis of a landmark paper that Lederberg wrote in 1978 titled, "Digital Communications and the Conduct of Science: The New Literacy."[49] Writing at a time when the ARPANET was transitioning to the Internet — a "network of networks" — and 12 years before "the World Wide Web" was invented, Lederberg predicted the huge benefits that would come from openly available databases, Web-based sources of information, and widespread electronic communication using what he called "EUGRAMS" (eu=true or genuine).

This would be the "new literacy," Lederberg said, and he anticipated that the technology would be available to the general public within the next few years. Email, he said, trumped the telephone because it was a more accurate and less disruptive form of communication.[50] He also foresaw junk email and the defenses that would be established against it.[51] There would soon be virtual bulletin boards and computer programs for creating and exchanging digital works of all kinds as he predicted. He also pointed to construction of a "world brain"[a] with the efficient refinement and sharing of human knowledge that resembles *Wikipedia* (founded 23 years later), or search engines like GOOGLE.[52]

Lederberg's views of new literacy had mixed reviews. Feigenbaum saw it as "amazingly far-sighted" and that Lederberg had acted as spokesperson for the entire national community.[53] Others

---

[a] H. G. Wells had elaborated his vision of the World Brain in 1938 — a new, free, synthetic, and authoritative World Encyclopaedia that could help the world's citizens make the best use of universal information resources and produce the best contributions to world peace.

were more circumspect. They saw computers as arcane instruments to be used only for very complex mathematical calculations. They were a little bit taken aback by Lederberg's insistence that the principal use of computers was going to be the handling of text and the communication of ideas.[54]

Chapter 19

The Crisis in Human Evolution

> The crisis is not only a political one, but not unrelatedly, a philosophical one. *Brave New World* comes close enough to the mark. The point is the imminent realization, in respect of many details, that man is a machine not only as a metaphysical abstraction but as an engineering program. Possessed of this kind of power, it is more important than ever that we have a clearer idea what man is for, and even before that, what man is.
>
> Joshua Lederberg to Jack Edman December 24, 1962[1]

During the 1960s, Lederberg turned to address the scientific community and the public on the sociopolitical and ethical issues that would soon arise from biological engineering and molecular biology.[2] The lack of forethought about biological engineering started to be clear to him in November 1962 when he participated in a four-day conference in London on "The Biological Future of Man." That meeting had a profound impact on him, leading him to question not only what it meant to be a scientist, but also what it meant to be human.[3]

The symposium attendees included some of the leading thinkers of the day: evolutionist Julian Huxley, Jacob Bronowski who is best known for his 1973 BBC television series *The Ascent of Man*, as well as several Nobel Prize laureates including Albert Szent-Gyorgyi, the biochemist and discoverer of vitamin C, Peter Medawar, the "father of organ transplantation," and Francis Crick, who had been informed a few weeks before the meeting that he was to share the Nobel Prize in medicine for his discoveries concerning the molecular structure of DNA.

When he received the invitation to the London symposium, Lederberg feared that the meeting might well be just "a frivolous whim," but he decided to take it seriously. Indeed, he would make it his own with an original presentation on the social and ethical issues around biological engineering, the "social dynamite" that he thought was likely to explode much sooner than most of his colleagues had faced up to.[4]

Lederberg suspected that much of the discussion would center on eugenics. After all, that was the usual way in which biologists had seen things with regards to the biological future of humans. Biologists had dreamed of improving the genetic composition of human society ever since Charles Darwin's cousin, Francis Galton, coined the term "eugenics" in 1883.[5] There were two general approaches to eugenics: "positive eugenics," which was aimed at increasing the population of humans deemed to be fit, and "negative eugenics," aimed at decreasing the reproduction of those deemed unfit. Eugenics acquired a particularly ugly reputation when, based on bogus conceptions of race and heredity, it was perverted and used for so-called "racial hygiene" and the mass murder of six million Jews, and millions of non-Jews, under the horrific programs of the Nazi regime. Despite the horrors of the holocaust and various other acts of genocide, the utopian hope of eugenics never died among biologists.

Lederberg, as per usual, took a much more active part in the discussions than he had originally anticipated. It began when he crossed swords with Julian Huxley, the grandson of Darwin's close friend Thomas Henry Huxley, and older brother of Aldous Huxley, the author of the classical and far-sighted dystopian novel *Brave New World* (1932). Julian was well known for forging evolutionary syntheses based on genetics and Darwinian theory. He was also an advocate of positive eugenics.[6]

Lederberg's quarrel with him occurred early in the symposium. Huxley argued that human biological evolution had been superseded by social evolution, or what he called "psychosocial evolution" in which "the struggle for existence has been replaced by what might be called the striving for fulfillment." Because evolution was the "most powerful and the most comprehensive idea that has ever arisen on earth," Huxley argued, "evolutionary humanism" ought to replace theistic religion and underlie all education. In humanity's quest for fulfillment, eugenics would relieve it "of great suffering and frustration, and would increase enjoyment and efficiency."[7] He thus advocated a so-called positive eugenics: selection for excellence based on artificial insemination by "preferred donors."[8]

Lederberg could not have disagreed more with Huxley, whose arguments he found to be puerile, illogical, and belying common sense. First, he noted that Huxley's argument about "evolutionary humanism" leading to "fulfillment" lacked a coherent definition of what it is to be human in the first place. Lederberg agreed that humans were outside of nature and that our social evolution has superseded organic evolution, but that implied to him that humans were no longer simply biological objects. What it is to be human had "evolved from substance to concept," as he put it, and eugenic aims would have nothing to do with that.[9]

Of course, neither Lederberg nor anyone else could disagree with the utopian idea of a more intelligent world population that might "keep itself from being blown up," but eugenics would not work just based on common sense.[10] First, little was known about human genetics — a single gene did not determine "intelligence" — and no one knew what undesirable traits went together with it anyway.[11] Second, attempting to achieve an improved human gene pool by, for instance, taxing undesirables who want children or applying the principles of animal husbandry to humans would entail the loss of humanity itself. It was absurd.

Not only were the old eugenics methods ineffective and unachievable in any socially acceptable way, but more precise methods for genetically modifying humans through the "integration of desired genes" were far in the future, as Lederberg saw it.[12] Moreover, scientists had no business deciding what was "good" for society. That was something that had to be decided by society itself.

In short, Huxley and the other eugenics advocates were simply wrong-headed. They also seemed blind to what was right in front of them. The immediate impact of biological engineering would not be on our heredity, but on our development — modifying humans through physiological and embryological alterations, with grafting of tissue and organ transplants. Those were the research problems around which Lederberg had organized his own genetics department at Stanford. As he put it metaphorically, "If we want a man without legs, we don't have to breed him, we can chop them off; if we want a man with a tail, we will find a way of grafting it on to him."[13]

Lederberg introduced the term "euphenics" at that London symposium. He offered it in contradistinction to eugenics to refer to control over individual development. He pointed to the future of organ transplantation and the industrial synthesis of hormones and other proteins that could play a role in prosthetic organs. So, he declared: "Man's control of our own development, 'euphenics,' changes the means and also the ends of eugenics, as have all the preceding cultural revolutions that have shaped the species: language, agriculture, political organizations, the physical technologies."[14] Euphenics meant improving health care and education, not genetics. To be human, as Lederberg saw it, was *to be a communicator*, and that meant "the preservation of our own future by the modernization of our techniques for efficient and free expression."[15]

The ethical issues of euphenics required deep political consideration. He could foresee the dehumanizing abuses of an unregulated market in human flesh for spare parts, and he called for governmental regulation and formal registration of all organ transplants from human and nonhuman sources. Lest he appear alarmist, he pointed to the negative effect of medical progress in the name of humanity: "the catastrophic leap in world population through the uncompensated control of early mortality." It was time to think deeply and anticipate the side effects of biological powers with the hope of developing social and technological antidotes.[16] So, he argued, first priority should not be given "to long range eugenic concerns," but rather to "gravely imminent issues of human numbers and phenotype: the allocation of intelligence, motivation and longevity."[17]

It might be hard for us now to understand how far-sighted Lederberg's comments were at the time, especially because organ transplantation is considered today to be one of the greatest achievements in medical science.[18] But none of this was apparent to those who attended the London Symposium; they were wedded to eugenics methods and thought that a progressive biological future was a matter of establishing "an effective kind of genetic betterment." Lederberg was astonished by their lack of insight.

"Whatever the value of this ethical discussion," he advised his colleagues at that meeting, "biologists could not decide these issues for the rest of the world." The general public and those in "the seats of high political power" needed to be informed of the technological possibilities of human modification.[19] He would aim to do something about that himself.

Lederberg's discussion of the ethical issues of biological engineering was inspired by Aldous Huxley's *Brave New World.* It is set in the World City State of London 600 years in the future where people are engineered through artificial wombs, children are bred into castes based on intelligence and labour, and citizens are indoctrinated and kept peaceful in part through a happiness-inducing drug called "soma."

Fig. 19.1. Jack Edelman

"Much of the time I felt that Huxley (Aldous) would have done better all by himself than the pack of us put together," Lederberg wrote to Jack Edman, his old friend from high school, after the London meeting.[20] "On a short term basis," he continued, "I would like to see more thoughtful insights on the euphenics problem... so that some sort of program can be put together for the sensible use of dangerous knowledge. Over the longer range, the definition of man does strike me as one of our crucial problems — I think we might not suffer too badly giving way to dolphins, computers and extraterrestrials, if this also helps us cope with the other monsters of our own creation."

Lederberg pointed to science fiction and other literature for insight on the problem of what constitutes a human. He had read Čapek's classic play *Rossum's Universal Robots* (1920), which were more like "Data" from *Star Trek* than what came to be called robots, and *War with the Newts* (1936), which explored the same theme from a different angle. He had also read Vercos's *You Shall Know Them* (1952) and he was sympathetic with Vercos's approach to understanding what it is be a member of humanity.[21]

The next month, Lederberg wrote to Aldous Huxley himself and sent along his paper on euphenics.

> One of your critics has remarked that "Brave New World" must have scared you even more than your readers. Having first been obliged to spend any real thought on the details of the future by the mission of a recent symposium in London, I found myself more deeply frightened by the immediacy of developments in human biology than I could have thought possible. It was impossible to budge in this field without bumping into a Huxley, Julian, Aldous or T. H.'s shade — on the whole, I found Aldous's biology the most persuasive...
>
> You know very well what a muddle science can make. But, however fearful the consequences of physical power, at least human nature

> has until now been left reasonably constant. The humanism we all seek will be far more elusive when we can no longer delineate the personality we hope to defend.[22]

Aldous Huxley visited Lederberg in Palo Alto in April 1963. He was an impressive figure, as Lederberg saw him, and though only 68 years old, he had cataracts and was nearly blind. He lived outside Los Angeles in his house at the top of Mulholland Drive beneath the Hollywood sign. They got along splendidly and spent a Sunday afternoon discussing "science and literature, nightingales and need for new poetry."[23]

Lederberg would prophesy about euphenics in a series of articles published in some of the most widely read scientific journals, beginning with *Nature* in 1963. The dramatic effects of biotechnology based on euphenics would be felt in five to twenty years, he said: one might soon be able increase intelligence by increasing the size of the human brain by prenatal or postnatal intervention; heart transplants were in the near future, and nonhuman species might be cultured for spare parts. Lest some readers assumed that the issues he raised belonged to a distant future, he explained the extraordinary rate of scientific growth. The amount of science conducted in one year by that time matched the total accumulation to the beginning of the 20th century, and there was as much scientific effort invested over the previous decade as there had been in all previous history.[24]

And further, he emphasized the need to define what "human" meant before one could protect humanity from its own technological creations:

> While man perfects the knowledge of his own mechanism, he also vitalizes machines on to a convergent evolutionary pathway. Genetics is

> rapidly becoming a corollary of information theory. As he thus evolves from substance to concept, is it the bond of genetics or of communication that qualifies 'man' for the aspirations of humanistic fulfillment, apart from the other robots born of human thought?[25]

Aldous Huxley died on November 22, 1963. Lederberg dedicated an essay titled "A Crisis in Evolution" to "the memory of his prophetic vision and artistic clarity." The sudden successes of molecular biology, he argued, "now demand an understanding of the nature and destiny of man that must be the principal intellectual tasks of the brave new world."[26] He had no doubt that what he called genetic alchemy, or "*algeny*" — "the chemical control of genomes" — would be realized one day and bring with it much social anxiety. And where did women fit into the eugenic aims? After all, he remarked, "most eugenic discussions have been over-whelmingly male-oriented, as in academic life." How could women's creativity and happiness be augmented by the chemical control of genomes such that XX is recombined with a set of male-oriented genes?[27] The best way to improve the human condition, he said, was to improve the lives of people, reduce poverty, and increase education and support for families.[28]

Lederberg's writings on euphenics culminated in 1973 with a masterful paper on "The Genetics of Human Nature." Eugenics, its social evils, and its dubious efficacy, he argued, had caused a paralyzing effect on thinking about human evolution. The important issues of the day, such as the atomic bomb, the human population explosion, and pesticides and pollution, were rightly at the forefront. But the forebodings about molecular biology — that it would soon give scientists the capability to directly alter human genomes — needed to be included in that list.

Fig. 19.2. Joshua Lederberg 1972.

Humans are outside of nature, he argued, and our future as a species is in our hands: "The biblical myth of the fall of man in Genesis was an expression of the preeminence of culture over nature, of the emergence of the superorganic beyond the organic, of the concept that man is a man-made species."[29] The responsibility of scientists was to teach people about the eventualities of biological engineering and the problems it would engender.[30] It was society's responsibility to decide both on moral criteria and on institutional remedies for these new technologies. The "insight of literary, political, social, economic and moral teaching" was needed to develop an understanding of these issues.[31]

Lederberg practiced what he preached. In 1968, he created an undergraduate course in human biology called "Man as Organism," which introduced Stanford students to the social and ethical implications of emerging biomedical knowledge. That course was the seed for a new and successful "Program in Human Biology" that involved collaborating colleagues from various specialties. The aim was to understand humans from biological, behavioral, social, and cultural perspectives.

The course developed from a unique social context in Stanford's medical school that fostered interdisciplinary interaction. Lederberg knew it had been courageous of President Sterling to move the clinical part of the medical school from San Francisco to Palo Alto. The clinicians in San Francisco certainly did not want the move to take place, but the scholar in Sterling realized the need to bring medicine in proximity to people from different fields.[32] The "Human Biology Program" was characteristic of that interdisciplinarity and of interdepartmental cooperation. There was an open spirit at Stanford then. Faculty wanted to cooperate and make it work. None of the big emerging issues of science and society fell neatly into one academic discipline or another. About a dozen faculty members from branches of the social sciences, biology, and chemistry joined together.[33]

"Human Biology" became one of the most popular programs on the Stanford campus. It developed into an undergraduate major, not because faculty wanted it to become a major, but because students demanded it. Those who took the first couple of courses had a sit-in in the president's courtyard, arguing that it should be a major and that more students ought to have the opportunity to take it.[34] Lederberg did not want the administrative responsibility of running the program, but he did help to make it permanent by insisting that the university find a way to get endowed chairs.

By that time, Lederberg had developed plans to bring these issues to a much larger audience beyond the university. He took on a weekly column in *The Washington Post* to educate politicians and policymakers.

Chapter 20 The Communicator

In the last five years or so, I have rather drastically changed my outlook on science... But while one is caught up in the process of discovery, one tends to overlook that the only realities for a human being are in his communications... My column-writing is a very satisfying vehicle for reaching these motives.

Joshua Lederberg to H. J. Muller, November 2, 1966

Lederberg had been transformed by the London Symposium, "The Biological Future of Man," in 1962. It broke him out of his solitude, his singular focus on developing personal knowledge, as he came to appreciate that knowledge itself was social and that communication was the object.[1] "Science is a social phenomenon, rather than one of private insight," he wrote to his friend Jack Edman:

"... Surely the common ground of all philosophy is verbal expression, and I am surprised (not to be aware) more has not been made of this as an alternative to eclecticism... What shall I call it — cultural idealism: that reality is not a private idea but the idea of my cultural tradition, from which I learn all the means of knowledge. Put another way, reality is the rule of communication within a culture..."[2]

With this new enthusiasm, he aimed to impart that cultural knowledge to the largest audience possible. His goal was to write a regular column in *The New York Times* or *The Washington Post*. He tried the *Times* but did not get very far. But he was successful when he approached the science editor of *The Washington Post*, Howard Simons, who had consulted with him on exobiology and NASA's "Man in Space" program.[3] Lederberg put together some sample pieces and Simons was persuaded to go with a weekly column on "Science and Man."[4]

Everything got rolling quickly in the summer of 1966. His column ran for six years (comprising some 250 pieces in total) and was syndicated in about a dozen newspapers around the world. He addressed various ethical, social, and political dimensions of science and science policy. Topics ran the gamut from NASA's space missions to matters of national security, environmental problems, human genetics and evolution, biological engineering, and warnings of emerging viral diseases and the risk of global pandemics.

Few scientists of Lederberg's stature had ever attempted to write on science and society for lay readers.[5] British biologist J. B. S. Haldane was one of them, and he bonded with Lederberg in their humanism and belief in the importance of extending the scientific viewpoint to promote rationalist thinking on ethical issues at the interface of science and society. Lederberg was particularly fond of Haldane's comment in 1929 that "The enemies of science alternately abuse its exponents for being deaf to moral considerations and for interfering in ethical problems which do not concern them. Both of these criticisms cannot be right."[6]

Some thought writing a weekly series was beneath someone like him.[7] But many more were appreciative. He was especially pleased when geneticist H. J. Muller wrote to him in the fall of 1966 about his first series of articles. "It is a remarkable series that has in my opinion out-Haldaned Haldane himself, being on the whole even sounder, partly because it is not so smart-Alecky. I hope it sets the beginning of a new kind of scientific writing for the public and the popular press."[8]

Lederberg treated his writing on science, society, and politics with the same scholarly rigor as he did any of his scientific papers. He read about 50 articles a week and travelled with a box of reprints. Theoretical physicist Sidney Drell, who cofounded the Center for International Security and Arms Control at Stanford, applauded Lederberg's rigorous and informed discussions. "There were a number of scientists who would get involved in public debates about something they didn't know what the hell they were talking about," he commented decades later, "but, that did not apply to Josh," he said, "You couldn't predict where he was going to go on anything."[9]

Certainly there were other scientists, including Lederberg's friend Carl Sagan, who became public figures through popularizing science. But Lederberg's writings were different. He didn't personalize them and they were not descriptions in the sense of "gee-whiz" popular science. His essays were arguments, a response to something someone had said, or a reaction to events in the tumultuous days of the Vietnam War and civil rights movement. He addressed social inequalities as well as public anxieties about scientific advances in molecular biology and reproductive biology, such as the birth control pill and abortion.

Yet, nothing was more urgent than addressing racist views and their supposed biological basis. Lederberg discussed the issue in various contexts, including education, social programs, and state laws. He began with the now famous case of Richard and Mildred Loving, who matched the stereotypes of white and "Negro." The couple were legally married under the laws of the district of Columbia, but when they requested the right to enter Virginia as husband and wife, they were sentenced to a year's imprisonment for marrying each other because their marriage had violated that state's Racial Integrity Act of 1924 — that was until that law was struck down by the US Supreme Court on June 12, 1967.

In a column titled "Virginia 'Biology' Based on Delusion" in October 1966, Lederberg argued that the concept of black and white races was biologically bogus because each "race" concealed great genetic diversity within it. A white person, according to Virginia law, was anyone who had no trace whatsoever of "any blood other than Caucasian" (the word "blood" taken allegorically to mean ancestral heredity).[10] But the law made no sense, as Lederberg saw it, because very few citizens would have knowledge of their ancestry to show that they did not have "any trace" of a disparate color. After all, he argued, the history of Eurasia is replete with explorations, invasions, and crusades, in which there was a scrambling of the major sets of genes. Superficial characteristics of skin color concealed great variation among individuals in personality and achievement.

The word "race" as used by the Census Bureau, the Civil Service Commission, the Office of Education Statistics, and other Federal agencies was also biologically meaningless, Lederberg argued. With his typical foresight, he suggested that "race" should be replaced by "ethnic affiliation," and the latter ought to be an individual's choice. The state, he concluded, should not take "any special interest in race as a factor in marriage any more than it does in linguistic background."

Of course, there was more to the myth of race than backward Virginia law. Race was as explosive a biopolitical issue in the United States during the 1960s as it is today. And there were those who opposed progressive efforts such as the civil rights movement, President Lyndon Johnson's "War on Poverty," and the "Head Start" program to improve education and nutrition for young children of low-income families. They claimed that such programs should cease because African Americans were intellectually inferior and eugenic programs should be promoted to limit their reproduction.

Perhaps the most notorious of these racists in the academic world was right at Lederberg's doorstep at Stanford: Nobel physicist William Shockley, one of the inventors of the transistor, which had revolutionized the electronics industry and ushered in the information age. He had moved to Mountain View near Stanford to start a silicon-based semi-conductor laboratory, and he is credited as being "the man who brought silicon to Silicon Valley." When Shockley was appointed Professor at Stanford in 1963, he turned his attention to something he knew nothing about: eugenics and race. Pointing to data indicating that black Americans tended to score lower in IQ tests, he argued "that the major cause of the American Negro's intellectual and social deficits is hereditary and racially genetic in origin and, thus, not remediable to a major degree by practical improvements in the environment."[11]

Lederberg considered Shockley to be an overt racist who was abusing his prestige in science to promulgate his bigotry. Lederberg confronted him publicly and forcefully. When Shockley's views were published in the *Stanford MD* in October 1966, Lederberg drafted a reply to it that was signed by the entire faculty in his genetics department condemning them as a "pseudo-scientific justification for class and race prejudice."[12] They doubted that genes played any important role in economic and social accomplishment by race, religion, class, or any other major social grouping. They advocated another prescription for social change: "work out the techniques of medical care, education and industrial and economic organizations that can create incentives and useful careers for the whole wonderful variety of human beings."[13] Their statement was subsequently reprinted in the Congressional record.

Lederberg countered Shockley's racist views again in his *Washington Post* column on January 27, 1968. Under the title, "Evidence Links Poor Diet to Forever-Stunted Minds," he discussed the evidence that good diet for preschool years is critical for brain development and that malnutrition was associated with poverty, superstition, illiteracy, crowding, and poor sanitation, all of which needed to be considered when assessing the relatively poor performance of African American children.

Race and IQ turned into a national debate when Berkeley psychologist Arthur Jensen published "How Much can we Boost IQ and Scholastic Achievement?" in *The Harvard Educational Review* in April 1969.[14] His opening sentence said it all: "Compensatory education has been tried and it apparently has failed." It failed, he argued, because IQ was largely a question of genes.

Shockley paid for reprints of Jensen's paper and had them distributed to every member of the National Academy of Sciences. He actively solicited support for his views from leading geneticists and evolutionary biologists, but found none.[15] He also called on the National Academy of Sciences to sponsor studies on the genetic basis for the inferred differences in intelligence of whites and blacks. When the Academy responded with a letter signed by many of its members condemning his views, Shockley claimed it was scientific censoring. Lederberg considered it to be nothing of the sort.[16]

IQ tests did not measure inherent mental ability, Lederberg argued. They were achievement tests. Every skill they measured had to be learned, whether that be vocabulary or arithmetic. There were no valid tests of "inherent mental ability."[17] Lederberg pointed to "hereditary deaf mutism" to illustrate his point. Up until the early 19th century, a child born deaf never learned to communicate in any ordinary way and thus was unable to participate in any of the learning that involves speech. Because such children would score almost zero on any IQ test, one would have to say that their "inherent mental ability" was also zero.[18]

Lederberg also wrote a critique of Jensen's claims that IQ was a measure of biological competence, and that compensatory education had failed and was futile anyway. In fact, Lederberg argued, the opposite was true: compensatory education had hardly ever been tried and would be difficult to implement with "increasingly bitter alienation of the races."[19] In this hostile context, he said, there was no doubt that "intelligence" was largely determined by genes, but only those that control the color of the skin.

Lederberg took a rationalist humanist approach to all biopolitical issues. He seldom argued on moral grounds and applied this understanding to other issues of the day, including a woman's right to an abortion and research on embryonic cells. As he wrote in a *Post* column in 1970, the laws restricting abortion "amount to compulsory pregnancy on a large scale, an intrusion on personal freedom and privacy as repugnant as compulsory abortion."[20] The main issue that divided him and those who opposed abortion and embryonic research was their philosophical position that the early fetus is indeed a human being. "My moral distress," he said, "is focused on the misery that will be perpetuated if we do not learn more about prenatal and genetic disease."[21]

Lederberg got into a little hot water though, when in 1967 he provoked his readers with suggestions of human cloning. Under the title "Unpredictable Variety Still Rules Human Reproduction," he explained that it might be possible one day to make a human clone by inserting the nucleus of a body cell into an egg with its nucleus removed. Cloning, he said, was no different than taking a cutting from a rose bush. It might make human reproduction more predictable and controllable. "Our reactions to such a fantasy will, of course, depend on just who is immortalized in this way — but if sexual reproduction were less familiar, we might make the same comment about that."[22]

Conservative American physician Leon Kass, flabbergasted with Lederberg's "callous and cavalier tone," replied in the *Post* the next month under the title "Genetic Tampering." He said that the "genetic manipulation in man raises fundamental and enormous questions — theological, moral, political." He also opposed embryonic research.[23] When Kass sent Lederberg a copy of his letter to the *Post*, Lederberg replied rhetorically: "Why is biological intervention called 'tampering' as it almost always is, and not education, good nutrition and so on?" He answered himself: "I think mainly because it is just less familiar? We ought to have a humane interest in all these ways of tampering with individual development!"[24]

Lederberg did not relent and, a week later in the *Post*, picked up on the point he'd raised with Kass, asking readers to keep a balanced perspective on how much human engineering was already practiced in child rearing and schooling.[25] The abuse of genetic manipulation by governments was the most tangible of rational fears in Lederberg's view. Already a report of the United Nations and Human Rights had warned that new biochemical studies may "soon lead to the temptation of trying to manipulate the human race in accordance to the preconceptions of those who are in control of the means of effective change."

Lederberg responded to that fear to emphasize again that such thought control was already occurring: "Every educator, publicist, or politician is of course equally dedicated to the same end," he said. Much of the fuss about "genetic engineering" was also based on "the exaggerated mystique about the importance of the genes compared with other, far more accessible influences on the development of human nature-like education, indoctrination, custom and other social institutions."[26] Genes no more made the individual than a house made a happy home.

No one could defend ideas about genetic engineering more skilfully, but he saw little value in human cloning; genetic engineering should be applied to amend body cells. Though Lederberg was one of the first to talk about human cloning as a practical matter, he left the issue by the early 1970s as he became involved with more pressing concerns of human biology.[27] Serious talk about cloning was essentially crying wolf when a tiger was already inside the walls. He was especially concerned with the potentially hazardous effects of chemicals in the environment such as lead and chlorine, some of which, though widely used, were not tested for their mutagenicity.[28] But above all else, he was troubled by the global threat of biowarfare, and he became deeply involved in national and international policy to try to prevent it.

Chapter 21

The Advocate

> The large scale deployment of infectious agents is a potential threat against the whole species: mutant forms of viruses could well develop that would spread over the earth's population for a new Black Death.
>
> Joshua Lederberg, 1966[1]

Senators and Congressmen, heads of federal funding agencies, and heads of private philanthropies read Lederberg's columns in *The Washington Post* as he addressed many of the central issues of the day — from the potential hazards of untested chemicals in the environment to the promise of biological engineering and the sociopolitical and ethical issues that it engendered. Lederberg hoped to affect government policy, and in that regard perhaps nothing he wrote about was more important than when, in 1966, he called for a treaty to end the development of biological weapons.

Research on chemical and biological weapons in the United States began during World War II. The Army had a well-known research facility at Fort Detrick, Maryland, and a testing station at Dugway, Utah. Lederberg had known about these places since his days at the University of Wisconsin in Madison.[2] The Geneva Protocol of 1925 prohibited the use of biological and chemical weapons, but the US had not signed it.[3]

Lederberg called for a new treaty that would go much further and ban not only the use of biological weapons, but also their development, production, and stockpiling. The US government heeded his call and asked him to take an active role in establishing one. He accepted gladly and worked with officials in the State Department, ambassadors, diplomats, and the World Health Organization. He was, in fact, the sole expert consultant during the negotiations in Geneva that led to the signing of a germ warfare treaty in 1972.

In the late 1950s and early 1960s, Lederberg was in deep despair about the possible annihilation of civilization by nuclear weapons, or indeed of the human species by biological warfare. There seemed to be little if anything that he or any other university professor could do to help prevent it. Physicists such as Einstein, and Lederberg's good friend Leo Szilard, who played a role in the effort to build the bomb, spent much of the rest of their lives in the effort to constrain it. Szilard wrote the letter that was signed by Albert Einstein and sent to President Franklin D. Roosevelt on August 2, 1939, urging that the United States develop the bomb before Germany did.[4] That letter precipitated the Manhattan Project. But once the bomb was created, Szilard, Einstein, and many others feared it would destroy civilization.

Scientists wrote letters and signed petitions urging world governments to stop nuclear testing and reach an international agreement to prevent the spread of the bomb. Nobel Prize winning chemist, Linus Pauling, who wrote petitions, gave lectures, and sent letters to President Kennedy, was awarded the Nobel Peace Prize in 1962 for his work. But Lederberg thought all of these efforts were, at best, simply useless.[5] In November 1957, Pauling sent Lederberg a petition with an appeal to scientists and citizens of the world to ban nuclear bomb testing.[6] Two thousand American scientists signed it. Lederberg declined. He thought that such a petition would have no effect behind the iron curtain, and "its propaganda effect" did "more harm than good."[7]

Lederberg considered international academic discussions on world peace to be similarly a waste of time. Philosopher Bertrand Russell and Physicist Joseph Rotblat founded the Pugwash Conferences (named after the village in Nova Scotia where they were first held in 1957), which brought together scholars to seek solutions to global security threats. When Russell invited Lederberg in March, 1959 to participate with an international group of scientists at a Pugwash Conference on biological and chemical warfare as weapons of mass destruction so as to inform governments of the risks to world security,[8] Lederberg declined, stating that he could not fit it into his schedule.[9]

When Rotblat sent Lederberg a summary statement and conclusions from that Pugwash Conference, recommending that biowarfare research be abolished and security restrictions on research be lifted, Lederberg replied rather haughtily: "I could not disagree with them as far as they go, but they seem to me so futile that I would not care to join publicly with them."[10] Given the poverty of trust and good faith between the West and the Soviet Union, he thought any such formal agreements would be wholly untrustworthy and unenforceable.

Lederberg's pessimistic attitude suddenly changed after the treaty on the non-proliferation of nuclear weapons was signed in 1968. He then became interested in the discussions at the Pugwash meetings and wrote to Rotblat asking for information.[11] He realized that he might be able to do some good after all when he acquired his weekly column in *The Washington Post* and when he teamed up with Harvard molecular biologist Matthew Meselson.

Meselson had been Linus Pauling's last postgraduate student. While Pauling wrote petitions against nuclear arms, Meselson took a stand against chemical and biological weapons. That happened, however, out of circumstance rather than design when a colleague and friend at Harvard arranged a summer job for him in the new Arms Control and Disarmament Agency of the State Department in 1963.[12] He was supposed to work on nuclear arms control, but because he had nothing to offer, he asked if he could work instead on chemical and biological warfare where he might be more useful.

Everyone had security clearance in the State Department then — even the janitors — for fear of communist infiltration, Meselson recalled decades later.[13] That meant that he was free to talk to anyone in the department. He learned from the CIA that they knew nothing about Soviet biowarfare programs. When he went to Fort Detrick and saw tanks full of anthrax, he asked why they were developing biological weapons. He was told that the agents of germ warfare were much cheaper to manufacture than nuclear weapons. That argument was insane in Meselson's view. It meant that anyone could acquire the technology, that germ warfare was "a poor man's nuke." He went into action to get the attention of the Oval Office to change policy.

Meselson and Lederberg worked together exchanging information about biological weapons from the summer of 1966 onward.[14] They were especially concerned about the United States' increased commitment to research on chemical and biological weapons in response to the war in Vietnam. The Defence Department's expenditures on those weapons had increased several fold since the 1950s, and it had begun "large-scale use of anti-crop and 'nonlethal' chemical weapons in Vietnam."

In September, Meselson sent Lederberg a draft of a petition that he wrote and addressed to President Lyndon Johnson, urging him to end the use of chemical weapons in Vietnam and begin the process of establishing a biological and chemical warfare treaty.[15] Some 5,000 scientists had signed the letter by the time it was released on September 19, 1966. But Lederberg, true to form, declined to sign it, just as he had when Pauling had asked him to sign his petition against nuclear weapons.

This time though, he had a plan of his own and, as it turned out, a particularly effective one. Chemical and biological warfare were not at the front line of national interest, but his weekly column for *The Washington Post* had just begun a few weeks earlier. He planned on using it to get the message out by commenting on Meselson's petition.[16]

First, he offered Meselson some advice on the text of his petition: "Shun the humanitarian side or anything that smacks of it, as diversionary," he said. "You would probably lose them on 'how awful it is.'" He suggested instead to focus more sharply on the arms-control aspect: the dangers of escalation and loss of control. "Scientists do have a special basis for comment since they can advise the President how easily such weapons might be aped by nth powers, and how they might get out of hand (at least for biological weapons)."[17]

Meselson sent him a new version of the petition three days before its release: "Josh — now is the time for us to urge you to write a piece for the *Post* on this matter. The weekend of the 25th would be an excellent time for its appearance."[18] Lederberg had intended to do just that, and so he did, adding an important twist of his own in terms of policy strategy. Whereas Meselson's petition called for a treaty on chemical and biological weapons, Lederberg called for a treaty exclusively on germ warfare.[19]

To be sure, Lederberg opposed chemical warfare, but the United States government was already committed to the use of chemical agents in the war in Vietnam and, as he saw it, it would be difficult to achieve a policy reversal. Defining and banning chemical weapons was going to take some time. So lumping chemical and biological warfare together was a strategic error. It was best to focus solely on banning biological weapons.

Lederberg explained to his *Post* readers that the United States was making extensive use of defoliating chemicals in Vietnam, not only against forest cover, but also against crops purportedly available to the Vietcong. Tear gas had also been used in military and occupation missions, and although the United States had vehemently denied using any biological weapons or any lethal use of chemical weapons, he explained that research on these weapons had continued since the beginning of World War II. "Biological warfare should be carefully set apart, particularly for the initiative in international negotiations for several reasons," he wrote:

> It is the most dubious of military weapons. Its effects in field use are most unpredictable, with respect to civilian casualties, and even retroactively, on the user.
>
> The large-scale deployment of infectious agents is a potential threat against the whole species: mutant forms of viruses could well develop that would spread over the earth's population for a new Black Death. Chemical weapons, however potent, at least do not produce equally or more virulent offspring!

As it turned out, Lederberg's call for a treaty solely on germ warfare resonated well in Washington. Meselson wrote to him the next month: "Did you notice that Senator Case referred to your editorial in *The Washington Post* in advocating what he described as a nonproliferation treaty for biological weapons?"[20] Still, it would be a few more years before the Government took real action.

In the meantime, Lederberg became active. He gave talks at scientific meetings arguing that biowarfare posed a threat to the human species, and he also fired off more warnings about germ warfare in the *Post*.[21] On August 31, 1968, under the title, "The

Infamous Black Death May Return to Haunt Us," he explained how the Black Death in Europe in the mid-14th century had started with returning Genoese traders from the Crimean seaport of Caffa on the Black Sea. Their ships were infested with rats that carried fleas, which in turn carried the bacterium *Yersinia pestis* that caused the bubonic plague.[22] "The deprivations of war, as in Vietnam," he said, "also foster new outbreaks of plagues."

Lederberg explained that mutant antibiotic resistant strains of the bacillus that spread the Black Death were easily cultured in the laboratory, and the resultant potential for germ warfare was obvious. He recommended that a defensive biological research program should develop such drug-resistant bacilli only to evaluate them as a military threat, and he was especially concerned with airborne viruses. He argued that the Soviets and the West should find "a common cause in protecting ourselves from a recurrence of pestilence." After all, germ warfare could add nothing to the strategic power of nuclear-armed states. The risk, he said, was its use by small nations or insurgent groups.

Lederberg revisited his warning about the return of the Black Death the following week when he coupled the fear of germ warfare with the outbreak of an unknown disease he called "the Marburg virus."[23] Under the title, "Mankind had a near Miss from a Mystery Pandemic," he told the story of events of the summer of 1967, when a group of at least eight "green monkeys" used in a laboratory for preparing vaccines contracted an extraordinarily contagious and deadly disease previously unknown to medical science. The disease infected at least 32 people and killed five of them in Marburg, Germany. It could have been an "epidemic of world shaking dimensions" dominating the earth "with a half a billion casualties" if it were not for sheer luck.

Science, Lederberg said, would have been ill equipped to stop such an outbreak from leading to a global pandemic. "Furthermore, a great deal of scientific ingenuity is dedicated to 'improving' such agents, the most suicidal of human enterprises today." For many in the US government, Lederberg's call was simply unwarranted fear mongering. The weaponization of the Marburg virus along with another hemorrhagic virus, Ebola, was in fact part of the Soviet Union's biological weapons program. But the West would only learn of that decades later (Chapter 27).[24]

Lederberg was persistent; he sent letters to senators he knew with his plea to distinguish policy with regard to biological and chemical weapons. He knew his way around Washington. He knew some of Kennedy's and some of Johnson's staff too. That summer, he wrote to Senator Walter Mondale as someone who might be willing to champion the cause to push for a biological warfare treaty and advised him to contact Meselson.[25]

Lederberg also advised the World Health Organization's Director of Science and Technology, Martin Kaplan, to separate biological warfare from chemical warfare as the best strategy to achieve an international agreement.[26] He explained that chemical warfare was caught up in protests against the war in Vietnam and that could cloud discussions about biological warfare, which had not been used. He and Kaplan would work together on a biowarfare treaty soon thereafter.

Government attitudes about biological and chemical warfare changed when an NBC television program on chemical and biological weapons aired in February 1969.[27] It revealed plans to test biological weapons on Baker Island in the South Pacific and showed scenes of laboratory experiments with rabbits and mice, along with images of dead sheep being bulldozed into huge pits in Utah's Skull Valley in March 1968 when a weapons test went wrong.

On November 25, 1969, Nixon unexpectedly renounced biological weaponry in any form. He pledged that the country would drop its work on the development of biological weapons, destroy existing stockpiles, and join in diplomatic efforts to negotiate a comprehensive biowarfare treaty.[28]

Meselson had also worked behind the scenes in Washington. In the Spring of 1969, the Secretary of Defense requested Henry Kissinger, as National Security Advisor, to review studies of their chemical and biological warfare programs. Meselson was Kissinger's source of information. Although Nixon paid little attention to the intelligence community or the State Department, he and Kissinger were close.[29] In their book, *Germs: Biological Weapons and America's Secret War*, journalist Judith Miller and coauthors asserted that Lederberg and Meselson's relationship was adversarial.[30] Yet, nothing could have been further from the truth. They were "brothers in arms" as they both saw it.[31] In fact, the month after Nixon's announcement, Lederberg nominated Meselson for a Nobel Peace Prize.[32]

Lederberg was recruited to advise the government as a scientific expert on the risks of germ warfare, just as he had hoped to be. Indeed, he would be called upon to provide expert testimony to the United Nations as it contemplated the need for a new treaty to outlaw not just the deployment of biowarfare, but also its development. That work began in December 1969 when he gave testimony on biological weapons before the United States' Committee on National Security Policy of the House Committee on Foreign Affairs.[33]

Offering expert advice to the government was not always well received by colleagues, students, or indeed some politicians. As protests over the war heated up, university scientists who worked for the government were sometimes verbally attacked for having "sold out" or for being "prostitutes."[34] Lederberg and Meselson took it from both sides of the political spectrum. While the left denounced them for selling out to a bad government, the right denounced them for their criticisms of government policies.

In March 1970, Charles Conrad, the Republican leader in the California State Legislature, attacked Lederberg and Meselson as being "antigovernment," "leftist," and "ill informed" for their work against chemical and biological warfare. "Scientists with strong political motives are twisting scientific facts to achieve ends detrimental to the nation," he said.[35] Lederberg certainly opposed the war in Vietnam and he thought the United States should take responsibility for its use of chemicals against foliage and crops. In May 1970, he signed a letter to President Nixon along with 39 Nobel Laureates calling for an end to the war.

Political leanings aside, however, Lederberg could hardly be called ill-informed when he offered "expert" advice, whether in his *Post* column or when giving testimony. His approach was scholarly and driven by logic and rational thought, not emotion or political ideology.[36] He maintained that same approach when giving testimony to government and international committees. As one of his colleagues put it, "That's why he was so valuable. He was smart and he knew how to keep politics out of what he was doing, when he was advising the government."[37]

The month after Conrad publicly besmirched him in the California State Legislature, Lederberg was invited to Geneva to work on the United Nations Disarmament Committee as advisor to the United States' Arms Control and Disarmament

Agency. Its aim was to form a biological weapons treaty which Lederberg had called for three years earlier.[38] He was now going to be the scientific expert in those very negotiations. He proudly wrote to his friend Salvador Luria in July 1970 about the invitation:

> They (the arms control agency) may well have focused on me for such a task, in part, because I have long adopted a position that distinguished between BW and CW on several grounds…
>
> At any rate, the US position on these matters is very close to my own, in terms of the priority with which various steps should be taken…[39]
>
> My own position on the matter is colored by the very grave fears that I have about virological warfare...[40]

Twenty-six nations were represented at the UN Conference when it convened in Geneva the following month. Lederberg was the only scientific expert. He wrote a memo detailing his experience during that week in August 1970.[41] He took on the role with the kind of scholarly approach he was known for, reading up on conference diplomacy and how to use scientific expertise in policy formation.[42] He attended seminars on public policy and international politics at Stanford's Political Science department, and he got personal tutorials from government officials. The Director of the Arms Control Agency discussed with him analogies with the Nuclear Disarmament Treaties and explained some of the complexities of decision-making in a democracy. The Ambassador to the United Nations also instructed him on the difference between policy aim and the processes of negotiating the contracts to solidify it.[43]

Lederberg explained to the diplomats in Geneva that he could not pretend to be an expert on chemical warfare; but he was certainly an expert on microbes and the potential risk that biological weapons posed to the world. He clarified that the Geneva Protocol only prohibited the use of biological and chemical weapons. A new treaty on biological weapons would also ban their development, production, and proliferation. Above all, he feared that genetic engineering would be used for germ warfare. He explained how bacteria could be modified by the direct manipulation of DNA "and result in a biological weapons race whose aim could well become the most efficient means for removing man from the planet."[44]

Lederberg's appearance as a scientific expert at the UN meeting and the information he supplied came as a surprise to many of the diplomats there. The Soviets in particular "made a rather haughty remark that we should not have experts at meetings like this…"[45] For his part, Lederberg was astonished at how little any of the diplomats seemed to know about biological warfare, and with the exception of the Soviets, there was great interest in what he had to say. Many of the diplomats in Geneva asked for a copy of the text of his talk. It was also presented to the United States House of Representatives the next month and subsequently published in the widely read magazine, *American Scientist*.[46]

Most of the questions that followed his talk in Geneva centered on why he distinguished a biological weapons ban from a chemical weapons ban.[47] He explained his reasoning based on precedent — that chemical weapons had already been used, but to his knowledge there had been no military use of germ warfare in modern war. The one event that they needed to avert would be "a biological Hiroshima" from which there might never be an effective return, he said.[48]

Giving separate consideration to biological weapons was agreed upon by Britain at the UN Conference, but the Soviet Union and its allies opposed it. They argued that banning germ warfare alone would appear to be legitimizing chemical warfare and undoing the work of the Geneva Protocol.

The impasse with the Soviets continued until a second UN Disarmament Conference was held in 1971. At that meeting, twenty-five delegates passed a draft Convention that prohibited the development, production, and stockpiling of both biological and toxin weapons. The United Nations General Assembly subsequently adopted the Convention by a vote of 110 to 0.[49]

Lederberg was elated when twenty-five governments ratified the new Biological Warfare Convention in the spring of 1972. "Mutual Bravos, on the BW treaty just signed," he wrote to Martin Kaplan at the World Health Organization. "I was pleased to be present at the signing ceremony in Washington. If there is a reasonable posture of global compliance, microbiologists can indeed feel liberated from the kind of burden that has plagued nuclear physics."[50]

The signing of a treaty, though a landmark, was certainly not going to be the end of the problem. Lederberg explained the situation in the *Stanford Journal of International Studies* in 1972: the United States had proceeded with destroying biological warfare stocks, and the former germ warfare facilities were being converted to work on environmental pollution, drug safety, and cancer research. However, the Soviets had yet to release comparable information on their biological warfare program.[51] Lederberg would be involved with biowarfare verification and attempts to establish transparency and trust with Soviet biologists throughout the remainder of the Cold War (Chapter 28).

Chapter 22

Scooped

What you may not know is that our own laboratory has also been working for the last several years along very similar lines but was "scooped"...

Joshua Lederberg to Martin Kaplan, WHO September, 1974[1]

No development in genetics offered more biomedical promise, in Lederberg's view, than cutting desired bits of DNA from one species and splicing them into the genome of another. Those techniques would make it possible to use bacteria to make insulin, human growth hormones, vaccines, and other products for treating and preventing disease. Stanford's medical school was where recombinant DNA techniques were first developed — most prominently in the biochemistry department. Techniques used to harness the enzymes that break and splice DNA were at the basis of this, and knowledge of nucleic acid biochemistry in Arthur Kornberg's department was unparalleled.[2]

Kornberg's department was critical to the origins of genetic engineering, but it was research in Lederberg's department that actually pointed to the so-called "miracle enzymes" used for cutting and splicing DNA pieces. Lederberg had a longstanding interest in recombinant DNA. In his Nobel lecture of 1959 he had discussed synthesizing DNA and modifying genes by inserting short pieces of DNA into them as a form of directed mutagenesis. The research program in his laboratory that was started when he arrived at Stanford aimed to understand the process of DNA replication and uptake in *Bacillus subtilis*.[3]

Lederberg wrote what some historians believe to be the first genetic engineering grant application to the National Science Foundation in 1967, in which he elaborated on its potential.[4] The ability to transplant human segments of DNA into a culture of microbes would raise the prospect of made-to-order enzymes for many biomedical purposes, he said.[5] He suspected that at least 25 percent of all medical problems had a genetic basis. Gene therapy was also a real possibility in the near future. Suitable viruses might be used to carry human genes to the body cells of patients with certain defects. He called it "a kind of transductional therapy for human genetic disease."[6] DNA injected into body cells by means of viruses would be a "genetic vaccine." The effects of "virogenic therapy" would not be passed on to succeeding generations any more than those of a polio or smallpox inoculation.[7]

One could also learn what was happening to our genes and what the function of those genes are during embryonic development by taking bits of DNA from different tissues and implanting them into a bacterial genome. "Suppose for example, you could get a DNA segment that codes for human pituitary growth hormone," Lederberg told a journalist, "transplant it into a bacterium, and then fire up this bacterium to produce the growth hormone. You'd have a full-scale human growth hormone factory operating in your laboratory."[8]

Walter Bodmer and Adayapalam Ganesan studied the molecular mechanism underlying the uptake of DNA transformations in *Bacillus subtilis*. The enzymatic steps involved in its integration remained unknown. They searched for enzymes that govern the way DNA is repaired naturally after it is torn apart by environmental assaults such as exposure to radiation. If Lederberg's group could find the enzyme that stitches the DNA molecule back together when it is severed, they would have a method that not only takes genes apart and puts them back together again, but that could also link up the genes of different species of organisms.

In 1967, Kornberg made the electrifying announcement that his lab had synthesized the DNA of a virus. They artificially duplicated a short 5,000 nucleotide-long single-stranded viral DNA and showed that when the single strand was converted to a double-stranded circular form, it could infect *E. coli* and behave like a natural virus.[9] It was the first successful synthesis of a biologically active virus, and as Kornberg told a journalist, after many years of trying, "the way was open to create novel DNA and genes."[10] The next year, Ganesan developed a system for replicating 50,000 to 100,000 nucleotide-long double-stranded DNA of *Baccilus subtilis.*[11]

Lederberg had moved to Stanford in 1959 partly because Kornberg was moving there. His department was adjacent to Kornberg's; they shared some facilities, held seminars together, and sometimes even held departmental meetings together.[12] But the cooperation was not deep and did not last. The close collaboration that Lederberg had hoped for never really came to be, and the relations between labs were not always cordial. Kornberg, at times, ran a closed shop. When Bodmer worked on the synthesis of DNA, he was not permitted to attend research seminars in the biochemistry department. Even when he was collaborating with a researcher there, he would have to wait outside and learn second hand what was said.[13]

Relations between the departments deteriorated further in the early 1970s when research on recombinant DNA methods in Lederberg's laboratory was carried out in implicit competition with Paul Berg's laboratory. Berg was one of the original six founding members of the biochemistry department, and he succeeded Kornberg as its Chair in 1969.

Recombinant DNA was all about finding the right enzyme for the job, and the so-called "miracle enzyme" turned out to be something called EcoR1. It is a restriction enzyme, produced as part of the defense system that bacteria use against viral invasions. It cuts up the DNA of invading viruses and prevents them from parasitizing the cell and reproducing within it. EcoRI became the key to cutting and splicing DNA to make desired recombinants. Italian chemist Vittorio Sgaramella, working in Lederberg's laboratory, discovered that EcoRI generates complementary DNA ends that can be joined.

Lederberg's old friend and collaborator Luca Cavalli at the University of Pavia recommended Sgaramella for Lederberg's program on "directed transformation" in 1970.[14] Cavalli himself was appointed professor of genetics at Stanford that year to replace Walter Bodmer, who had returned to England to take up a position at Oxford. Sgaramella arrived the year after. He had been at the forefront of research on the laboratory synthesis of DNA, having collaborated with Nobel Prize-winning molecular biologist Gorbind Khorana and his group at MIT in a landmark achievement in synthesizing the first complete wholly artificial or synthetic gene.[15]

Sgaramella's aim at Stanford was to extend that research so as to cut and join natural DNA.[16] At that time, EcoRI had just been discovered in Herbert Boyer's laboratory at the University of California in San Francisco. Berg and his graduate student, John Morrow, showed that EcoRI could cut DNA reliably at a unique site.[17] They thought the enzyme made blunt ends when it cut DNA, but when Sgaramella tested it in Lederberg's lab in the spring of 1972, he found that not to be the case. Rather than making flush ends, the enzyme actually cut in a staggered way that left short complementary single-stranded DNA ends, which could easily be joined by another known enzyme called *E. coli* ligase.

Sgaramella's observation that EcoRI generates "sticky ends" led to a flurry of research on the EcoRI enzyme in Palo Alto and San Francisco that year. Berg's student, Janet Mertz, and Berg's colleague, Ronald Davis, confirmed Sgaramella's observations, and Boyer and his collaborators quickly conducted experiments it showed the exact nucleotide sequence of that staggered end. EcoR1 always makes a signature cut with the overhanging sequence AATT, which has a complementary palindromic sequence TTAA.

EcoRI would become a most powerful tool for recombinant DNA research.[18] Because the enzyme made identical cuts in any piece of DNA from any organism, fragments would naturally be complementary and thus join without further biochemical treatment. In 1973, Stanley Cohen at Stanford's Department of Medicine, together with Boyer and their students, devised ingenious methods using EcoR1 to successfully insert a foreign gene into a bacterium where it could be cloned.[19]

On the surface, this episode might appear to be a case of good cooperation between Lederberg's laboratory and the biochemistry department — precisely what Lederberg had hoped for when he moved to Stanford. Unfortunately, that was far from it. A priority dispute emerged at the time of publication over who actually discovered that EcoRI cuts in a staggered way to create complementary DNA ends. Sgaramella was bluntly cut out.

Sgaramella thought that his priority in the discovery was protected when Berg gave him pages of the draft manuscript on EcoR1 in which Sgaramella's work was acknowledged.[20] But that acknowledgment was deleted before the paper was published. Sgaramella was devastated. He later showed Lederberg and Stanley Cohen those draft pages that Berg had given him recognizing his priority before he published his paper.[21] They could only shake their heads. Berg shared the Nobel Prize for his work on recombinant DNA in 1980 while Sgaramella was removed from all historical accounts of the discovery emanating from Berg's department.[22] But in 1995 when Cohen readdressed the question of who first discovered the ability of EcoRI to generate complementary DNA ends that can be joined by the enzyme *E. coli* ligase, he credited Sgaramella.[23]

The collaborations that Lederberg had originally hoped for with the biochemistry department were all but gone for him personally by 1970. Sgamarella recalled that during his four years at Stanford, Lederberg did not attend biochemistry seminars or interact much at all with Kornberg or Berg.[24] He was involved in various activities — in artificial intelligence and the search for life on Mars while being heavily committed to national science policy and establishing a biological warfare treaty.

But it was not just a matter of Lederberg's preoccupations elsewhere. His approach to inserting genes was flawed and did not lend itself to the level of precision required. His strategic error was in his insistence on focusing on the uptake of DNA by the bacterium as a whole.[25] He looked upon cells in their entirety, as an interacting system, and had an aversion to "grinding them up and extracting their individual components, which is how biochemistry made most of its advances."[26]

Cohen and Boyer bested him in 1973 when they learned to insert genes into bacteria by conducting more refined experiments at the right level of complexity. They devised a method of inserting a spliced piece of DNA within the small donut-shaped loops of *E. coli* plasmids.[27] So, as Lederberg commented decades later, Cohen "just had much sharper insight into what the appropriate methodology was and I remember very vividly meeting him in the parking lot out in front of the Medical School and his telling me the experiments he had just concocted and my saying, 'Well, there's five years' work down the drain.'"[28]

After Cohen and Boyer's success in inserting genes into *E. coli* plasmids, several groups of scientists immediately planned to do likewise. This was the connection between molecular biology and medical application that Lederberg had hoped for in cloning genes: the prospect of making medical products such as insulin and antibiotics as well as new methods for investigating cancer and the pathogenicity of viruses. But all this work was put on hold as a major public storm broke out over the risks of recombinant DNA technology. Lederberg and Berg would find themselves on opposite sides of the controversy.

Chapter 23 Prometheus Unbound?

> I am pleased that you brought up the question of Berg's committee on DNA biohazards. To explain part of my position… I think there really is great danger of the whole matter getting seriously out of hand and encumbering important research not only in molecular biology and in cancer biology but also in the study of pathogenic viruses generally… One point I have not been able to get across to Dr Berg is the extent to which every hospital is potentially a source of the difficulties that equal or exceed the "new" fabrications that may result from splicing DNA.
>
> Joshua Lederberg to Martin Kaplan,
> World Health Organization, September 23, 1974[1]

A great public furor over recombinant DNA research broke out in the 1970s. While Lederberg saw its great promise for the study of cancer and pathogenic viruses, and for such medical products as vaccines and antibiotics, others pointed to environmental risk, public safety, and the ethics of "tampering with life" just for the curiosity of scientists.[2] For the first time, some declared, the "species-barrier" that prevents genetic crosses between unrelated organisms had been broken. There was a range of reactions to this: some called for government guidelines restricting certain kinds of experiments, others wanted all gene-splicing experiments stopped, and still others were concerned about the partnership between university and industry that biotechnology seemed to entail.[3]

Public outcry was sparked by the actions of Paul Berg, Lederberg's neighbor in the biochemistry department down the hall from his genetics department at Stanford. Berg had planned a rather hazardous experiment to insert a cancer-causing virus, SV40, into *E. coli*. He only changed his mind about this after one of his students, Janet Mertz, who was taking a course on tumor viruses at Cold Spring Harbor, told one of her course directors about the plan. Alarmed by the prospect of anyone inserting a cancer-causing virus into *E. coli* (which inhabits the human intestinal tract), her course director phoned Berg to object.

Berg was a molecular biologist, not a medical microbiologist. He had not considered the potential biohazards that would arise from his proposed experiment. Shocked by what he was about to do, he radically changed course. He chaired a National Academy of Sciences committee meeting with ten other scientists who were attending a Gairdner Conference on nucleic acid. The committee sounded the alarm in a letter to *Science* in July 1974 citing two major risks: (1) the creation of "novel types of infectious DNA elements whose biological properties cannot be completely predicted in advance," and (2) that the "new DNA elements introduced into *E. coli* might possibly become widely disseminated among human, bacterial, plant, or animal populations with unpredictable effects."[4]

They thus called for a moratorium on any experiments that might involve inserting genes for antibiotic resistance, toxin formation, or the insertion of cancer-causing viruses into bacteria strains until the potential hazards were well evaluated or better methods had been developed to prevent their spread. They requested that the National Institutes of Health establish an advisory committee to devise national guidelines to be followed, and that an international meeting be convened in the next year to discuss ways to deal with potential biohazards.

Lederberg was dismayed by Berg's committee's letter in *Science*. Its central propositions were reasonable enough, but he took umbrage at their statement "that self-regulation in science has little or no precedent." This claim, as he saw it, was irresponsible and erroneous. He knew of several precedents where risks were carefully examined before research was undertaken. In fact, he himself had been among the leaders when it came to such precautions as interplanetary quarantine (Chapter 11), from which international regulations were instituted for specific standards of risk with respect to planetary contamination.[5] He had also been among the first to see the ethical issues that might arise from organ transplantation when he called for national and international regulations to prevent the emergence of markets for human flesh (Chapter 20).

To be sure, Lederberg recognized that the risk of some gene-splicing experiments should be assessed. In his view, though, laboratory safety for gene-splicing was not really much different than that for handling pathogenic organisms. As a biochemist, Berg simply did not appreciate that many of the safety practices were already commonplace among researchers who handled infectious disease. Lederberg tried (unsuccessfully) to get the point across to Berg that every hospital laboratory is potentially a source of difficulties that equal or exceed the risks of any new constructions that could result from splicing DNA.

Berg's careless statement about unregulated science, as Lederberg saw it, threatened to open a whole can of worms. His fear was that unwarranted government regulation could be imposed with "what if?" scenarios that might arise from ignorance and exaggerated risks. The result would be government bureaucracy that could smother the advancement of research and forestall potential medical benefits.

Lederberg had been down this road several years before. After all, anxiety over the potential of genetic engineering had been a simmering issue since the emergence of molecular biology. He had tried to dampen some of these fears in his *Washington Post* columns, but he was especially worried about an emerging anti-science movement arising from the anti-establishment counterculture with its distrust of government, fuelled at that time by an unwarranted and unpopular war in Vietnam. Indeed, while Lederberg was concerned about government control *of* science, others were concerned about government control *by* science.

At a 1969 symposium on "Public Considerations in Genetic Technology," the fear was expressed that molecular biology might soon have "the ability to create new forms of life that might lead to the generation of a new form of subhuman species to perform man's menial tasks in a condition of slavery." Much of that anxiety, Lederberg countered in the *Post*, was based on "the exaggerated mystique about the importance of the genes" compared with other far more accessible influences on human nature, such as education, indoctrination, customs, and other social institutions. Depriving children of proper nutrition and education would easily make intellectually inferior humans, as would depriving a woman of an abortion when the fetus is infected with German measles.[6]

Lederberg replied similarly to the famous playwright Edward Albee, who feared "mind control in the future." Those concerned with mind control, he said, "should look hard now — at television, advertisements, when they promote cigarettes."[7] Fears about governments' control through genetic engineering peaked again when Harvard molecular biologists Jon Beckwith and collaborators announced in *Nature* in 1969 that they had isolated a few genes in *E. coli.* Beckwith held a press conference stating that "the use by the Government is the thing that frightens us." He suggested that a malevolent government could genetically engineer a more docile public or a "super race."[8]

Lederberg countered Beckwith's "what if" scenarios with a rejoinder in his *Post* column stating that "Biology could add little to the Hitlerian Repertoire: exterminatoria, forcible impregnation and sterilization, drugs and thought control."[9] Beckwith's fears were misguided as Lederberg saw it. Genetic engineering would be used on bacteria and viruses for medical purposes. Slavery under an authoritarian regime would much more likely result from moral, social, military, and political disasters than from science. The greatest potential for government abuse was biological warfare, which he had been involved with preventing from inside government channels.

Beckwith was part of the radical science movement that accompanied the civil rights movement, rise of feminism, distrust of government and state control, opposition to the war in Vietnam, and the overarching threat of nuclear war which coincided with the emergence of genetic engineering.[10] He was a member of the organization "Science for the People," which also decried the commercial link between universities and private industry.

While critics of the emerging biotechnology deplored university links with industry, Lederberg encouraged them. He himself was on the board of Syntex Laboratories and on one of the first biotech startups in Silicon Valley, CETUS, and he was a consultant on antibiotics research for the British pharmaceutical company Bristol Laboratories.[11]

In a letter which is as relevant to issues of today as it was when it first appeared, Lederberg addressed "the anti-science counterculture" in *The New York Times* in January 1970.[12] The scientific potency of molecular biology in the 1960s, he argued, needed "to converge with the environmental conservation movement of the 1970s," and he advocated the precautionary principle with regard to environmental matters — better safe than sorry. He forewarned that chronic alterations to the environment "may be too subtle for simple common-sense observations. One method we must avoid is to wait for rigorous proof of widespread damage before we react with precautionary measures."[13] After all, he argued, "the planet could be committed to the ash heap before such a fate was proven by the standards of laboratory experimentation."[14]

In 1973, the year before Berg's biohazard committee's letter appeared in *Science*, Lederberg spent a year as a scholar in the Program of Science, Technology and Society (STS) at the Center for Advanced Study in the Behavioral Sciences at Stanford. There he met Harriet Zuckerman and Robert Merton, with whom he would become close and lifelong friends. Anti-science was a primary concern of the STS program, "affecting the morale and funding of scientists, students' choice of careers, and the whole fabric of a technology-dependent civilization."[15]

When put in historical context, it is easy to understand Lederberg's concern about Berg's letter which spoke of "unregulated science" and called for a moratorium. That letter, Lederberg believed, was fated to open a Pandora's box of inflammatory public concerns, provide fodder for the anti-science movement, and further erode the professional autonomy of science. In the meantime, he did his best to try and mitigate what he foresaw as imminent public outcry, hyperbole, and overreaching regulation by government funding agencies.

In August 1974, a month after the Berg letter was published, Lederberg wrote to the Director of the National Institutes of Health, Robert Stone, about its "careless statement" about unregulated science.[16] "The fact is," he said "that scientists are so law-abiding and generally so responsive to social sanctions that many of these concerns have been internalized, are accepted informally without great fuss and question, or are part of the overall legal framework in which every citizen operates."[17] Lederberg agreed that some new institutional safeguards might be required, but he feared unnecessarily stringent safeguards that might impede research and which, once introduced, might be difficult to change even if they were eventually shown to be unnecessary. Lederberg was upfront with Berg and sent him a copy of his letter to Stone.[18]

Lederberg also explained his position to Martin Kaplan, the Director of the World Health Organization's Office of Research Promotion and Development, which would also soon issue guidelines. Kaplan knew Lederberg well as a member of the WHO's advisory committee for medical research. They had also worked closely on matters of tropical disease and on the biowarfare treatise of 1972 (Chapter 23). "The central propositions that started this whole matter are reasonable enough," he told Kaplan in September 1974, "but it is hard for me to communicate the madness that has then inflated the proceedings."[19] Lederberg recommended that instead of acting in isolation, the WHO ought to participate in an international symposium that had been called for in Berg's committee letter.

Asilomar

That meeting would be held six months later in February 1975 over a four-day period at the Asilomar Conference Center in Pacific Grove, 90 miles south of Stanford.[20] There were 140 participants, including scientists, physicians, lawyers, journalists, and government officials.[21] As Stanley Cohen recalled, the discussions were "very Talmudic" in that they focused on hazards that weren't known to exist and relied on a lot of assumptions and speculation.[22]

Lederberg showed up with a box of reprints to read and a statement he had written for distribution at the conference.[23] He explained many of the medical benefits to be derived from recombinant DNA research, including the search for a cure for cancer and for genetic diseases.[24] He then pointed to the hypothetical risks of the comparable or greater threats that were left unpoliced in other areas of infectious disease research. "Science", he said, "is in grave danger of losing its intellectual autonomy if the pursuit of certain kinds of knowledge is judged to be illegitimate."[25]

On the last day of the conference, Berg asked for a vote on the statement that certain experiments should not be done at that time, and that other experiments be conducted with appropriate physical and biological barriers to contain the newly created hybrid-like organisms.[26] The great majority agreed to it. Lederberg, Cohen, and James Watson would not sign it. Lederberg explained in *The New York Times* that he was "wholeheartedly in support of the spirit and intentions of the Conference Report," but was unwilling to put his name to "a document that left many important questions for future determination, and whose tone seemed to invite… bureaucratic rigidity."[27]

The conference organizing committee submitted its final report to the National Academy of Sciences on April 29, 1975. The National Institutes of Health assumed responsibility for translating its recommendations into detailed guidelines for research. Before its guidelines appeared, and fearing that they could be rigid and impede research, Lederberg continued with his writing on benefits and risks.

Where such a hazard is reasonably predictable, such as introducing potentially cancer-causing genes into common bacteria, Lederberg agreed that laboratory containment precautions should be taken akin to those appropriate for known pathogens. But the best strategy for safety, he argued, was actually in the biotechnology itself: engineer plasmids and bacteria "so that they have little chance of surviving outside the laboratory." In fact, in the long run, this is a safer procedure than relying upon the uncertainty of human compliance with fixed rules and regulations."[28]

He also addressed critics who falsely claimed that moving genes from one species to another through gene splicing was unnatural. He pointed out that the concept of species as closed gene pools did not readily apply to the bacterial world because different "species" could readily exchange genetic fragments naturally through conjugation and viral vectors, and they can also acquire DNA from the environment.[29] In other words, nature already carried out "experiments" in DNA splicing.[30]

All of this was to no avail. The NIH Guidelines issued in June 1976 identified experiments that were not to be performed and different levels of containment based on the potential hazard of other experiments. It also created an oversight structure: forms for grant applicants as well as a system of reporting and assessing. But then, as Lederberg saw it, the situation only got worse. The resultant so-called "recombinant DNA wars" quickly spread to involve local governments, citizens, public organizations, and politicians across the country.

Some critics of the NIH guidelines said they did not go far enough and relied solely on scientists' professional responsibility; others wanted the research to be stopped altogether as they talked of the creation of monsters, "playing God," and upsetting the order of nature with fears of untold dangers for human kind.[31]

Scientists advocating gene-splicing research like Lederberg did were made out to be self-serving and venal. It was all just idle curiosity: scientists wanted to do it simply because they could. These notions weren't mere distortions coming from an uneducated public; they also came from scientists themselves. The well-known biochemist Edwin Chargaff said: "Have we the right to counteract, irreversibly, the evolutionary wisdom of millions of years, in order to satisfy the ambition and the curiosity of a few scientists?"[32]

In August 1976, under the headline "New Strains of Life or Death" on the front cover of *The New York Times Sunday Magazine*, recombinant DNA was touted as a potential holocaust.[33] That article was introduced into the Congressional record the following month.[34] The possible impact on the environment and public health as well as on ethical and social issues captured congressional attention and resulted in over a dozen legislative proposals aimed at imposing new restrictions on the development and use of the technology.[35] In the summer of 1977, the city council of Cambridge, Massachusetts created the Cambridge Biohazards Committee to conduct site visits and review containment measures for all proposed experiments, with the aim of protecting residents from potential health risks.

Lederberg's former student, Norton Zinder, who had attended the Asilomar Conference, was dumbfounded by all of this. He wrote to Lederberg in June 1977 recalling his own warnings that going public could result in exaggerated claims and overregulation. Zinder said that he wanted to "blow the lid off" the whole issue. Lederberg replied that nothing constructive would come from doing that and suggested that "the painstaking and patient effort at educating the relevant people in Congress" seemed like the only hope. "'I told you so' is the furthest thing from my mind, Norton," he replied. "What has been happening just hurts too much to even think of dealing with it at that level."[36]

Lederberg's warning that regulations could become codified and difficult to undo was borne out. The biomedical research community organized an extensive effort to present new evidence that recombinant DNA was not hazardous.[37] For example, when Berg and his group expressed their concern about potential risks, they thought that such experiments would be conducted on a wide variety of *E. coli* strains, and that some of these could establish themselves in nature or in the intestinal tracts of humans and animals. But such experiments were not carried out on a wide variety of *E. coli* strains. *E. coli* K-12, which Lederberg had used in his first genetic experiments in 1946, and its plasmids and phages became the model system. That organism was thought to be severely "crippled," having spent fifty years or so in laboratories, and therefore was not likely to survive for long in the competitive environment of the human alimentary tract. All the risk assessment experiments subsequently done on it confirmed that it was indeed safe.[38,39]

The NIH would change its guidelines in 1979 — after three years. Many experiments were judged to be harmless, and containment for other kinds of experiments was reduced.[40] Guidelines for recombinant DNA research were not only relaxed further in the early 1980s, but also coupled with an increase in both corporate and federal funding through the NIH. Biotechnology companies mushroomed, facilitated by patent and tax policies that encouraged joint university-industry research, and permitted universities to patent the results of research funded by taxpayers through the federal government.[41] Some academic scientists were forming their own companies or agreeing to license their findings to established biotech firms.

Lederberg was only glad to see the advancement of biotechnology for medicine. The first successful implant of human insulin genes in *E. coli* was reported in January 1979, and mass production of insulin in bacteria began in 1982 at Genentech, which Herbert Boyer, who had pioneered gene-splicing, cofounded as the first major biotech company in Silicon Valley.[42] The production of interferon for protection against pathogens and tumors by inserting human genes into *E. coli* was carried out in 1980, and mass production by the Swedish firm Biogene followed shortly thereafter.[43] Gene therapy research for treating genetic diseases began in the next decade at the National Institutes of Health.

Many years later, looking back on the Asilomar conference and assessing what had happened, Lederberg and Berg remained polarized. Berg saw the Asilomar meetings as having changed the world of science policy, that scientists were able to gain the public's trust because of the media's presence, and that the meeting came to a consensus view.[44] Lederberg had the opposite take when he commented on the Asilomar meeting in light of the importance of recombinant DNA methods for research on viruses during the HIV pandemic: "The sensational publicity given ca. 1975 to the hypothetical hazards of recombinant DNA research ignited public fears and regulatory reactions that boded ill for the opportunity to continue research on methodology of the most crucial importance for the understanding of virus infection." Some participants at the Asilomar Conference, he lamented, viewed such research as if it were "an idle diversion for the amusement of scientists; therefore what harm in an indefinite moratorium?"[45]

In the meantime, another conflict with Berg arose when Lederberg was called upon to be president of one of the most formidable biomedical research institutions in the country.

Chapter

24 Unexpected Turnabout

> … I was thinking of accepting a high administrative position back East and I talked to him a lot and he was sort of discouraging. And then by God he went to do it himself. And I reminded him, why'd you do that? But when I'd go to New York, my wife and I would get together with him and Marguerite. He's just one of these people who's very smart, a wonderful friend.
>
> Sidney Drell[1]

Lederberg was getting restless by the late 1970s and decided to leave Stanford and move back to New York City to become President of the Rockefeller University.[2] In some ways, he was coming full circle. The Rockefeller Institute, as it was then called, was where Avery and colleagues had produced evidence in 1944 that DNA might be the basis of the gene. That work had sent Lederberg down the path to bacterial genetics and earned him a Nobel Prize. Since then, as we have seen, that path branched off in many directions.

Yet Lederberg's appointment as president of the Rockefeller came as quite a surprise to his colleagues. It seemed so out of character because he did not enjoy administration. He was a reluctant institution builder, doing only what he had to do to further his research interests. In fact, he had considered moving back to New York in 1967. At that time, his *alma mater*, Columbia University, was looking for an outstanding biologist with administrative skills to head a newly formed Department of Biology.[3] Lederberg said he would want his title to be "professor of biological sciences and the human condition" and to have his AI research continued there.[4] He certainly had no stomach for being head of a large department. His only interest in the position at all was that he feared that the excellent faculty in his genetics department would soon be getting irresistible offers elsewhere.

That indeed happened by the time he accepted the offer from the Rockefeller University in 1977. Walter Bodmer had long since moved to Oxford. Eric Shooter had left the genetics department to chair his own department of neurobiology in 1974. Elliott Levinthal, who had run the NASA instrumentation laboratory, would soon leave Stanford to work for the Defence Advanced Research Agency for three years.

Lederberg's closest friend, David Hamburg, who founded the psychiatry department, had moved to Washington to become President of the Institute of Medicine at the National Academy of Sciences in 1975. Lederberg grieved when Hamburg left Stanford.[5] They had worked closely together in developing the human biology program and studies of science, technology, and society. They also had a shared interest in public policy and arms control.[6] Lederberg admired Hamburg and his research on the biosocial roots of aggression.[7] A photo of Hamburg was one of three framed images in Lederberg's office at the Rockefeller University. (The other two were his Diploma from the Institute of Radio Engineers in 1961 and his certificate as a fellow of the American Academy of Microbiology.[8])

Anyway, Lederberg had long moved away from laboratory research *per se*. His grant applications in molecular biology were being returned with requests for more specific details on research focus and expectations. He had long ceased having any close

Fig. 24.1. David Hamburg, undated The Joshua Lederberg Papers.

relations with Arthur Kornberg's biochemistry department, which had enticed him to move to Stanford in the first place. Lederberg's relations with Paul Berg, who now headed the biochemistry department, were stressed by issues of priority in the development of recombinant DNA techniques and later by the controversy over its regulation (Chapters 23 and 24).

Lederberg had written a letter to the National Institutes of Health to counteract what he saw as Berg's alarmist views about the risks of gene splicing, arguing that bureaucratic regulations could have a stifling effect on the progress of science. Lederberg never made his dispute with Berg a personal issue though, and he was transparent in his actions. Berg, however, did make it personal, and he went behind Lederberg's back to undermine him. He went so far as to try to engineer a coup to prevent Lederberg from becoming president of Rockefeller University.

Before Lederberg was formally hired at the end of 1977, Berg summoned two Rockefeller faculty members, James Darnell and Norton Zinder, to attend an "urgent meeting" at Yale where Berg was then visiting. The two men drove up to New Haven to see what was so pressing and met with a small group of scientists. Berg advised them to do whatever they could to keep Lederberg from being hired as president.[9] His appointment would be a disaster, in Berg's opinion, because he was too disorganized to run a biomedical research institution the size of the Rockefeller, his genetics department had been unproductive, and he did not make good use of the funds he received for laboratory space.

Biochemists such as Berg seemed to have little appreciation for, or perhaps even knowledge of what Lederberg had accomplished. They certainly failed to recognize how he had advanced and shaped the careers of colleagues in his department. They may have also known little about NASA's instrumentation laboratory and its contributions to NASA's missions, and no doubt knew little if anything of Lederberg's work on artificial intelligence applied to science. Nor did they seem to care much about his having introduced computing systems to the medical faculty at Stanford.

Yet, members of Lederberg's own department and his collaborators admired him for his leadership roles. Although he was a disorganized person by his own admission, he was ironically also noted among those who worked with him for his administrative prowess — both intellectually and institutionally — in raising millions of dollars from private donors.[10] "Those of us who shared project management responsibilities with him," Edward Feigenbaum who worked with him on AI, "were in awe of the unerring accuracy with which he was able to effect clean cuts through thickets of political, financial and institutional complexities to lead us through to desired goals."[11]

Lederberg perhaps never learned about the secret meeting that Berg had organized against his appointment as president of the Rockefeller. But several months after arriving at Rockefeller University, he described some of his Stanford's genetics department's accomplishments to Stanley Cohen who succeeded him as its head, and who needed to understand the department's institutional funding. His version departed dramatically from the account Berg had given to Darnel and Zinder.

Lederberg explained that the space that his genetics department occupied in the medical center was funded with millions of dollars secured from the Rockefeller Foundation, the Kennedy Foundation, NASA, and by matching funds from the NIH and "the larger share of those dollars" contributed to the general construction of the entire building. "I do not believe there can be any doubt about the extent to which the activities of the Genetics Department have fully corresponded to the expectations laid upon us by our donors," he told Cohen.[12]

Both Darnell and Zinder were circumspect when they heard what Berg had to say. Darnell long knew of Lederberg. He had first met him at the National Institute of Allergy and Infectious Diseases in Bethesda during the fall of 1958. Darnell was a postdoctoral student then, working on the primordial molecular biology of the polio virus. Lederberg talked with him for ten or fifteen minutes and then explained to him what he was doing. "Before I could get out of my mouth what I was doing with polio, Josh already knew what the answer would be… So that was my introduction to Joshua Lederberg."[13] Darnell didn't heed Berg's advice. He did nothing overtly to abort Lederberg's appointment as president of Rockefeller University, and he didn't think Norton really wanted to, either.[14]

Zinder's own relations with Lederberg were complicated. He had been Lederberg's first graduate student at Wisconsin, where he was put on the path to discover the role of viruses in genetic transduction in bacteria (Chapter 8). He had been at the Rockefeller for 25 years by the time Lederberg arrived as president, and had made a name for himself for his discovery that some bacterial viruses carried RNA instead of DNA.

Although Lederberg was only three years older, he was still something of a father figure to Zinder.[15] Their relationship was complicated further in that Zinder resented not having had a share of Lederberg's Nobel Prize. As Darnell put it, "Josh's brilliance over shone everyone around him including Norton — a *very* bright man."[16] Zinder was a well-respected voice at Rockefeller University and he had mixed feelings about Lederberg becoming president at the institution in which he had worked for so long. He warned the trustees of the university that although Lederberg was profound, encyclopedic, and had huge capacity for work, he would be an indifferent administrator.[17] That was a pretty accurate depiction.

Lederberg's sudden decision to become President of the Rockefeller University was also a shock to his friends, Robert Merton and Harriet Zuckerman at Columbia University in New York. This was not just because of his distaste for administration, but because neither he nor his wife, Marguerite, were unabashedly urban, so they might find New York a trying place to live. "Bob and I have been thinking about you and wondering where you were in the decision process," Zuckerman wrote to him in November 1977.[18]

If the decision to move to New York surprised his friends, it stunned his wife who felt definitely excluded from any decision-making process. She was completely opposed to moving to New York as she was just starting her career in psychiatry and enjoyed her life in Palo Alto. She loved California and felt that the weight of history had been lifted when she moved there from the East Coast in 1962. Life in California helped her to forget her traumatic early years during World War II when her family was hiding from the Nazis.[19] She was married with a one year-old child when she first met Lederberg, who was 12 years her senior but looked even older. He reminded her of her father.

Marguerite was born in Paris in 1937, the only daughter of Polish Jewish parents who had escaped to France in 1936 and become citizens. Her father, Adam Stein, was a medical doctor who was well-loved by his community in Paris.[20] A good violinist and chess player, he wrote poetry and translated French poetry into Polish. He believed in Soviet-style communism and was part of the Paris Commune until the Molotov-Ribbentrop Pact (also known as the Hitler-Stalin pact) of August 1939.[21] Devastated by that event, he spoke out against it among his cadres and had to leave Paris.

The family moved to Jarnac, a small village in Southwestern France near Cognac. Their Judaism was not a secret then. In 1939, her father was sent to Angers, on the Maine River at the edge of the Loire Valley, to work in a hospital for Polish volunteers who got hurt during the war. The nuns who ran the hospital "adored him," their Jewish Doctor, and he was a soft-spoken man, not athletic looking but "extremely kind."[22]

When France capitulated to the Germans in June 1940, her father returned to Jarnac and started practicing again. As his practice grew, he hired a young apprentice doctor who lost his sense of humanity and wanted to turn her father in and take over the business when the German army occupied the area. When the mayor of the town got information that the Nazis were on the way to arrest Marguerite's family, he immediately plotted to get them out of town.

They fled to Saint-Victurnien, a small French village where the *Maquis* Resistance fighters greeted them. They lived in a place above a barn for much of the time, and Marguerite was looked after by anyone who could be trusted in one of the safe houses in town when her father and mother were with the Resistance fighters. Her parents brought her up as if she were Roman Catholic, sending her to church and to catechism. It was taken for granted by the nuns that she was a little Catholic girl.

Life in Saint-Victurnien was hell towards the end. Undernourished when not starved, she felt very alone and a burden to her parents. Saint-Victurnien was near Oradour-sur-Glane where the Nazi SS army massacred 642 people in 1944. Her father killed himself that year, but Marguerite learned of this only when she visited Jarnac in 1969 to meet some of those who had helped her family escape.

Marguerite and her mother moved to New York after the war. She was a precocious young girl and entered Bryn Mawr in Pennsylvania when she was just 16. Three years later, she entered medical school. She had tried for Harvard but was turned down due to limited quotas for women, Jews, and New Yorkers, and she fit all three categories. She was accepted at Yale in 1957, however, and received her medical degree in 1961. There were only three women then in a program of 60 to 65 students. She had entered medical school when she was just 19 and graduated at 23.

In her junior year, she married her first husband, Thomas Kirsch, who was also in medical school at Yale. He was from California, and they moved to Palo Alto in 1962 where they took residency at Stanford. Hers was in pediatrics while his was in psychiatry: he began training as a Jungian analyst at the C. G. Jung Institute in San Francisco. His parents were great disciples of Carl Jung and they cofounded the C. G. Jung Institute in Los Angeles. Marguerite loved her new family.

After her first child was born in 1964, she worked part time as a research fellow on a project she had devised to correlate the signs and symptoms of childhood diseases with their diagnoses. She had learned that Joshua not only had a computer in a medical center, which was virtually unheard of at the time, but that the computer could read and pick out words. She thought it might provide an efficient way to assess how many possible diagnoses there are for various symptoms.[23] She went to see Lederberg in 1965; that's when she first met him — over a computer, learning to program and enter data.

He was 40. She was 28, but she saw in him some of the characteristics that she remembered in her own father. Though he was considerably older than her, she found his "mind irresistible." It was the first time, she said, "I had met someone who could even begin to match my childhood memories of my father ... I could see that he was very good. And I could see that his moral values were very good."[24] She left her husband and his family to marry Lederberg in 1968 and, with her young son David, began a new life. A few years later, she began to specialize in psychiatry at the department that their good friend David Hamburg had founded. The Lederbergs' daughter, Annie, was born in 1974.

Marguerite was shocked that Josh had told her nothing of his intention to move to New York until the final negotiations. She had no interest in moving there, and her 12-year-old son, David, was equally upset because he had not been included in the decision to move.[25] It was neither an easy move nor a happy one. Still, Marguerite was resigned to offer Josh full support and give it her all in New York City. She started searching for a psychiatric position.

They travelled to New York for a reception with David Rockefeller and other board members when Josh was formally offered the position at the end of 1977, at which time Marguerite had set up interviews for a job of her own. She accepted a position as one of 24 doctors at the Memorial Sloan Kettering Cancer Center just west of the Rockefeller on York Avenue, where she was asked to help start a psychiatry department, the first of its kind in a cancer hospital.

In January 1978, *The New York Times* heralded Lederberg as the new President of the Rockefeller University, and he and Marguerite were swept into another world where they wined and dined with David Rockefeller and his spouse, Margaret, and other distinguished New Yorkers. They moved into the President's house, a beautiful mid-century building at the northeast corner of the campus on York Avenue. The house was huge, and dark; it came with a staff of four: a secretary for the missus, a cook, and two maids. At first, Marguerite was miserable not knowing what to do. But the Lederbergs became friends with the outgoing president, Frederick Seitz, and his spouse Elizabeth. He was "a little distant and waspy New Englandish," she recalled, but Elizabeth was softer and nicer, and helped her a lot.[26]

Chapter 25

The Rockefeller

> Speaking personally, when I came to the Rockefeller I missed the breadth of interdisciplinary perspectives with which I associated daily at Stanford... The city, beyond our walls, has much to offer in compensation; but there are still perils of scientific parochialism — perhaps especially in an institution with the high reputation in our specialty that the Rockefeller has enjoyed for many years. (You should peruse the enclosed brochure to see how specialized it is!).
>
> Joshua Lederberg to Arthur Norburg, March 21, 1980[1]

Lederberg often found being the fifth president of the Rockefeller University rather trying. It was hardly the pocket-sized department that he had headed at Stanford. With 60 laboratory heads and some 200 scientists, it was on a wholly different scale. But the challenge was not just in managing the scientists. He was hired at a time when the institution was in the midst of an identity crisis and under severe financial stress.

The Rockefeller had undergone several transformations since its founding in 1901. It was created as the Rockefeller Institute for Medical Research by John D. Rockefeller Jr, whose father had founded the University of Chicago two years earlier. At first, it was simply an organization that awarded grants to study public health problems, such as bacterial contamination in New York's milk supply. Then in 1906, laboratories were opened on the site of a former farm at York Avenue and 66th Street.

It was the United States' first biomedical institute, and it essentially followed the model of France's Pasteur Institute, founded in 1888, and Germany's Robert Koch Institute, founded three years later. The Rockefeller Institute's Hospital opened in 1910 as the nation's first center for clinical research, which linked laboratory investigations with bedside observation and treatment. The Rockefeller Foundation, albeit established as a separate entity in 1913, was closely connected to the Institute.

By the time Lederberg was hired, the former biomedical research institute had lost its way in the eyes of its Board of Directors. The problem arose when Detlev Bronk, formerly president of Johns Hopkins University, became president of the Institute in 1953. He instituted big changes and expanded the institution's scope to include all of the natural sciences, social sciences, and humanities. He also established a predoctoral degree program. The result was that the Rockefeller Institute was rebranded as the Rockefeller University in 1965.

Bronk retired three years later and was succeeded by Fredrick Seitz, who had no biological or medical background at all; he was, rather, a solid-state physicist. Soon after he was hired, the university entered a period of severe financial stringency. It had been living entirely off its endowment, and its Board of Directors called for both a balancing of budgets and a renewed concentration on "the life sciences and related natural sciences."

All of this resulted in an institutional crisis by the fall of 1976 when Seitz issued statements to six philosophy professors advising them to seek jobs elsewhere. Academic tenure was threatened, and the story was covered on the front page of *The New York Times*. Although the institution had nine scientists who had won the Nobel Prize in the previous five years, a malaise had crept in and, in the eyes of some of its leading scientists, the place had been "coasting" and the science was "not as broad or quite as deep as it could be."[2]

Seitz was accused of bungling and mismanagement, and he offered to resign in the summer of 1976. The Chair of the Rockefeller's Board of Trustees informed the *Times* that a successor would not be able to take over until the mid-1978, to which the journalist quipped, "Anyone who pacifies spirits will be a candidate for the next Nobel Prize — for peace."[3] This was the context into which Lederberg arrived.

Patrick Haggerty, an engineer and CEO of Texas Instruments who had succeeded David Rockefeller as Chair of the Board of Trustees, convinced Lederberg that in refocusing the institute, he would have a lot of say as to who would be hired.[4] That was because unlike in typical universities, there were no departments at the Rockefeller, just laboratories — each was headed by one individual and comprised of a team of junior faculty and technicians.[5] When a laboratory head retired, there was no onus on the institution to keep up their particular line of research. That meant that Lederberg could be creative and hire the best from any field. Lederberg would also have much to say about promotions as well as laboratory space. Very little vote taking was done at the Rockefeller, and the faculty expected justice and fairhandedness from the president. It was "like a family sort of organization," as Lederberg saw it, with him at the helm.[6]

The word in the halls before Lederberg arrived was that he was really into computing and was going to modernize the place — which, in fact, he did from laboratory to office.[7] He reemphasized the institution's traditional strength in biomedicine, expanding it to include research on heart disease, cancer, mental and neurological illness, as well as infectious diseases, while retaining the institution's commitment to basic research. And although the Rockefeller had been designated as a university, Lederberg knew it was anything

Fig. 25.1. *back row* Fred Seitz, Joshua Lederberg, *front* Betty Seitz, Marguerite Lederberg, Pat Haggerty, circa 1978.

but; he put a plaque outside the university gates stating: "Formerly the Rockefeller Institute for Medical Research."[8]

For the institution to grow, however, Lederberg knew he needed to get heavily involved in fundraising. About half of the overall budget of the university came from federal funds. Much of the other half was committed to salaries for the heads of the sixty laboratories, and for the physical plant and so forth. Those basic operational funds were wanting when Lederberg took over as president. They urgently needed a million and a half dollars per annum just to break even, not for growth, and they were dipping into endowment funds.[9] A year and a half after taking on the position, Lederberg thought he would be fortunate just to keep the ship afloat over the next five years.

After all, unlike other private universities such as Stanford or Harvard, the Rockefeller did not have alumni donors. It also didn't have a large student body as it accepted only about 20 to 25 students a year, and each one took several years to complete their doctorate.[10] The university had established an MD-PhD program in collaboration with Cornell's Graduate School of Medicine, adjacent to it, in 1974. Later, the Sloan-Kettering Institute across the street also became a partner. Nonetheless, by the time Lederberg was hired in 1978, the institution only ever had about 400 graduates.

Therefore, raising funds fell largely on Lederberg's shoulders. At Stanford, he carried the administrative responsibility for acquiring a large amount of funding from government agencies and private foundations for operations in his genetics department. He was the proxy for a number of faculty members and he had 1.2 million dollars to his name there.[11] But innovation at the Rockefeller required 9 to 10 million dollars every single year.[12]

As the figurehead of the institution, Lederberg worked hard to gain the trust of private donors. He could easily convey what was going on in the university and how it related to the future of medicine, and he could do it in a way that was accessible to the intelligent, generally informed, lay public. It was his business to know everyone's research in a variety of fields and explain that to donors. This was his strong suit — as it turned out, no one could do it better nor with as much pleasure.[13] In fact, he was known to be able describe to donors what laboratory heads were doing better than they themselves could.[14]

Lederberg was not good at cocktail party-style fund-raising, and he did not relish that. But when he was one-on-one with a donor, he was extraordinarily effective. Donors saw the quintessential intellectual in his office with floor to ceiling books, a pile of offprints and manuscripts of scientific papers on a table in the middle of the room, and perhaps a half-eaten donut on display. None of this was contrived like it was for, say, James Watson, who, as Director of the Cold Spring Harbor Laboratories, would deliberately untie a shoe lace and mess up his hair a little so as to concoct an image when he met with potential donors. Donors grasped Lederberg's intellect the moment he spoke, quietly and with the precision of a verbal surgeon.[15]

Lederberg had long said that he enjoyed the library as much as he did the laboratory. At the Rockefeller, he developed a broad perspective of what was going on in many scientific institutions. "I wouldn't enjoy my job at all if it didn't give me both the time and necessity of continuing to touch on the scientific literature," he told a reporter in 1979.[16] He saw himself as "a kind of information central."[17] His colleagues at Rockefeller often gave him their manuscripts to read for comment prior to publication. When he scribbled memos on their manuscripts pointing them to other information that might be relevant, he was careful to suggest that they might know of that information anyway to ensure that no offense was taken.

Although he was not a chummy sort of person with faculty, everyone appreciated Lederberg's intellect. As his colleague James Darnel commented, "His fantastic memory and his eclectic interests were obvious to anyone who sat down with him, and his ability to understand anything you were telling him was instantaneous."[18] He enjoyed the intellectual aspects of his new job, knowing full well what research was going on and where it was headed.

Lederberg had little interest in the day-to-day operations, however. As his colleague Jesse Ausubel put it, "Josh wanted to be chief rabbi; he wanted to be an oracle; he had the vision thing in spades, but he didn't really want to be CEO."[19] In fact, Rodney Nichols was startled when Lederberg offered him the position of Provost of the Rockefeller when they first started working together as administrators. Nichols had worked as vice president for Seitz since 1970, but he thought it would be inappropriate for him to have that title with a Nobel Laureate president. They opted for executive vice president. Nichols was in charge of administrative details. Lederberg also hired Richard Young as vice president in 1979. He had worked at NASA, helped start the exobiology program there, and was the chief scientist for the Viking mission to Mars.

The Rockefeller's administration was meagre, especially when compared to universities today which have become so administratively top heavy. Even as late as 1987, it still lacked some of the basic structures familiar to universities. When Cynthia Greenleaf was hired that year as associate vice president of administration, given her years of experience as Assistant Provost of the University of Chicago, Lederberg told her he was glad that she had come from a university to help imbue similar values and structures at the Rockefeller.[20]

Nothing was more striking about the Rockefeller's organization when Lederberg arrived than its lack of a tenure track structure. The organization was very hierarchical: only laboratory heads had tenure and all junior faculty reported to them. Lederberg helped to change that structure so that young researchers could be promoted into tenured positions as Associate Professors.[21] He was also successful in hiring outstanding scientists.

Nichols recalled that Lederberg had a deliberate policy of hiring women scientists, and if some got off to a slow start, he urged that they be left alone for a while.[22] He had long become interested in the writings of psychoanalyst Clara Thompson on "the psychology of women and the problems of womanhood." While recognizing different drives in women, she articulated the way in which society distorted their biological differences, which in turn, had a discriminating and suppressing effect on women.

In 1969, Lederberg read a volume of Thompson's selected papers on "interpersonal psychoanalysis" and found them "to be a revelation." He recommended to the editor that a paperback edition be published for millions of other readers. "I do not have to emphasize the current ferment in problems of education of women," he said, "and their and our continued frustrations in seeking to create the social institutions that can begin to be as well adapted to their needs as we demand they be to ours."[23] This "special treatment" may rankle 21st century thinkers, but at the time, it was seen as progressive and to the benefit of women's advancement in male-dominated fields particularly.

Still, as Lederberg's good friend David Hamburg, who also served on the board, knew, he was uncomfortable in the role of university president: "It was a bigger and tougher administrative role than he had anticipated. But he learned a lot about finance, and he could exchange with financial guys on the board very well."[24] Lederberg's main difficulty as university president was not with the Board, as Hamburg recalled, but rather with some faculty members' egos.

Despite Lederberg's lack of interest in administrative details and the difficulties he encountered, the position sparked another renaissance for him: a major virtue of living in New York City was its proximity to Washington where he could more actively participate — often on weekly and biweekly commutes — in advising the government on national and international issues of war and peace.

Lederberg's interest in national policy was certainly in line with that of other presidents of the university, including Bronk, who had been president of the National Academy of Sciences for 12 years, and Seitz, who succeeded Bronk as president of the Academy in 1962. Seitz had also served on the Academy's Space Science Board (of which Lederberg was a founding member) in the late 1950s and 1960s.[25]

And so Lederberg's position as President of the Rockefeller certainly enhanced his involvement in policy matters. He chaired President Jimmy Carter's Cancer Panel from 1979 to 1981, advised the Department of Defence at a time when there was concern that the United States was losing its lead in military-related technology, and chaired Congress's Science and Technology Assessment Advisory Council. But his heaviest commitment in Washington was as a member of the National Academy of Sciences' Committee for International Security and Arms Control, working to prevent the development of biowarfare.

Chapter 26 Advice to Presidents

> ... Right now might be a rather propitious time for a fairly general study of expert advice to the government. It would not be confined at all to the White House, although the latter should probably be the centerpiece.
>
> Joshua Lederberg to David Hamburg, November 5, 1986

In 1986, when Lederberg suggested to his good friend, David Hamburg, then president of the Carnegie Corporation of New York, that he should establish a study commission on scientific advice to the government, Hamburg was receptive. Andrew Carnegie had established the foundation as a philanthropic fund in 1911 to support education programs in the United States, and later the world. The Carnegie Commission on "Science Technology and Government" began in 1988 as a five-year study aimed at assessing how different government departments might make use of science, and how scientists themselves could be more effective as advisory experts on science and technology issues. The study was highly influential; many of its recommendations have had a lasting effect. "I don't think I would have done it if Josh did not want it so much," Hamburg recalled.[1]

Lederberg had worked as an advisor for a wide range of government agencies, so he had a good understanding of how scientific advice filters through political decisions. His main concern at the time he recommended the study to Hamburg was with the responsibility of the scientist experts. There were two key aspects to this, as he saw it: confidentiality and conflict of interest. Scientific advice was certainly not a matter of giving "objective" advice, of course, because being knowledgeable on a subject meant, by definition, having an interest in it one way or another. What was important for Lederberg was that one declares conflict of interest before agreeing to take on any governmental advisory role. He had been very conscientious about this himself.

When he advised the World Health Organization on issues regarding regulation of genetic engineering, for example, he made it clear that he was a strong promoter of its use for medical products. In February 1976, he refused to be the chair of a recombinant DNA committee for the World Health Organization because of his advocacy for its exploitation. "I am happy to be a member," he wrote to WHO's Director, Martin Kaplan, "but I think it would be better for the chairman to be less intensely interested in the outcome of this special field from the standpoint of his own personal interests."[2]

He made the same declaration of potential conflict of interest ten years later when he was asked to be on a panel on biotechnology policy for the US Congress's Office for Technological Assessment (OTA). By that time, commercial interest in genetic engineering had grown, and he had been deeply involved as an advisor to various biotech companies. Lederberg had worked with the OTA on several committees, including one that focused on determining cancer risks from chemicals in the environment. But when he was asked to be involved in policy on genetically engineered organisms, he was careful to explain that he had "many roles in scientific and industrial development of biotechnology."[3]

Sometimes though, such experience was precisely what the government wanted. "I appreciate your sensitivity to potential conflicts of interest," the OTA's director, John Gibbons, replied, "but there's no need to worry that you might be an industrial wolf in academic sheep's clothing. In fact, the breadth of your contacts beyond academia is an asset for our purposes. Our panels are chosen to draw on all sectors of society and are meant to elicit the full scope of viewpoints and constituencies. It's a kind of Archimedes principle applied to policy analysis!"[4]

Conflict of interest was not the main issue, however, that sparked Lederberg's suggestion that Hamburg begin a Carnegie Commission study on scientists' advice to the government. There was another, more thorny issue — whether members of government committees ought to be bound to confidentiality. This was a simmering concern that involved a decade-long controversy over the demise of the "President's Scientific Advisory Committee" (PSAC) in the 1970s.

Presidents had always recognized the need for lawyers, economists, military officers, and other professionals in their inner circle, but not so scientists. President Harry Truman created the position of scientific advisor in 1951, and President Dwight Eisenhower greatly strengthened that position in response to the Sputnik Satellites. The PSAC persisted under the Kennedy and Johnson administrations, but Richard Nixon eliminated it in 1973.[5]

Nixon abolished it for two reasons. One was that he allied the scientific community with opposition to the war in Vietnam. But the main reason stemmed from a PSAC report on a billion-dollar proposal for development of the supersonic transport (SST) aircraft, which was designed to travel at speeds greater than that of sound. That program would have been the most intense commitment any government had made to technological innovation for nonmilitary purposes. The central justification for it was to shorten the time of transoceanic flights, but the SST had plenty of drawbacks: noise generated at takeoff, sonic booms during flight, the high cost of development, and the greater expense of travel compared to other types of planes. For these reasons, the PSAC recommended cancelling the program.

But Nixon chose to ignore its advice and announced that he was going ahead with the project anyway. When he refused to make any portion of his scientific advisory's study public, Richard Garwin, who headed its SST panel, spoke out; he ultimately testified before Congress and the project was cancelled shortly thereafter. A brilliant physicist, known for the development of the first hydrogen bomb, Garwin had worked at IBM since 1952 and had long been a consultant advising government on matters of technology and defense, especially nuclear weapons policy.

Garwin thought that because Nixon was so corrupt and had misled the nation on this and other matters, his membership on the PSAC did not bind him to silence under these circumstances. But when he publicly challenged Nixon's decision on the SST, Nixon decided to eliminate the PSAC altogether.[6]

Garwin's actions resulted in an unresolved controversy. Should such a scientific advisor have to forfeit his or her right to speak publicly in this context? Lederberg thought so; and he thought that what Garwin had done was wrong. Garwin and Lederberg knew each other well. They had met at Stanford and worked together on the JASON defense advisory group on a project concerning the refueling of jet fighters in the air.[7] They saw each other more frequently, and often socially, after Lederberg moved to New York.

Lederberg had the highest regard for Garwin and he fully agreed with his committee's position against the SST project. He himself had written mockingly about it in one of his *Washington Post* columns: "Why not establish Washington as the center of the sonic boom evaluation tests so that Congress can knowingly bear the responsibility for what will be inflicted on the whole population?"[8] Still, he disagreed with Garwin's actions in breaking the confidence of the president. "And that is a delicate question," he said two decades later. "As a matter of political science theory, I'm a little bit more on Nixon's side, because I think that's the only thing that can work. If you can't sustain the confidence of the President, how are you going to have a board of advisors?"[9]

Lederberg applied the same rule to himself as a government advisor. He had the highest security clearance, perhaps higher than any biologist of his generation. And he believed that his role as scientific advisor meant forfeiting his right to speak publicly on policy issues. As one of his colleagues remarked, "he was always the soul of discretion."[10] He had declined an offer to join a PSAC panel in 1967 because of his weekly column on science and public affairs for *The Washington Post*. How could he write on an issue on which he was holding information in confidence for PSAC? "I look forward to the time when I might be able to give a different answer to such a suggestion," he wrote to Philip Handler, "but it would really be out of the question for me right now."[11]

There was no scientific advisory committee for the administrations of Gerald Ford or Jimmy Carter. But there was a call by leading scientists to revive it in 1986 during Ronald Reagan's second term in office. Fatefully, that recommendation occurred in a context that was similar to that which led Nixon to eliminate the committee in the first place. This time, there was deep concern among socially responsible scientists over the Strategic Defense Initiative (SDI), a proposed missile defense system, parts of which would be based in space to protect the United States from attack by shooting down nuclear missiles. President Reagan had long championed its development to defend the United States against what he called "the evil empire" of the Soviet Union.

He announced the SDI in March 1983, shocking critics who dubbed it "Star Wars".

Two Nobel Prize-winning physicists, Hans Bethe and John Bardeen, publicly voiced opposition to Reagan's SDI program and proposed restoring PSAC in a *New York Times'* op. ed. titled, "Back to Science Advisors."[12] Bethe had worked on the development of the hydrogen bomb at Los Alamos during the war, and he was a leading voice behind the signing of the 1963 Partial Test Ban Treaty prohibiting atmospheric tests of nuclear weapons. He was head of the board of the *Bulletin of the Atomic Scientists*, a magazine established by Manhattan Project scientists who subsequently devoted themselves to ending the nuclear arms race.

Lederberg had been a subscriber to the *Atomic Bulletin* almost since its inception in 1945. When Bethe invited him to join its board in 1981, he did not hesitate in accepting.[13] "Although many of the articles in the *Bulletin* strike me as leaning to excessive trust in the goodness of the world," he replied, "the quality of information and debate on subjects of crucial importance would be greatly impoverished without the *Bulletin*."[14]

But when Bethe and Bardeen called for reinstituting the PSAC, Lederberg immediately challenged them. Of course, he was all for renewing it. That was certainly not the issue. The problem was that Bethe and Bardeen had failed to address the main reason it had been abolished 15 years earlier: the matter of confidentiality. So when Bethe and Bardeen called for reinstating the PSAC, Lederberg replied in *The New York Times*:

> Mr Bethe and Mr Bardeen are among the country's most respected scientists. They have, as is entirely appropriate, publicly voiced their own convictions critical of the Strategic Defense Initiative ("Star Wars"). It would have helped clarify the case for a Presidential Science Advisory

> Committee if they had also articulated how members of such a committee might have the same privilege.[15]

This was the context in which Lederberg suggested that Hamburg establish a Carnegie Commission study on "expert science advice to the government." What surprised him was that he got little flak from his colleagues for his response to Bardeen and Bethe in *The New York Times*. On the contrary, he received some encouragement to pursue the issue. As he explained to Hamburg on November 5, 1986: If the results of such a study "were to come out in mid or late 1988 it could have some salutary influence on the way the next administration is organized. No matter which party comes in at that time I am sure there will be a receptivity to new thinking that has not prevailed during the current regimes."[16]

While it was ignited by the call to reestablish the PSAC, the Carnegie Commission on Science Technology and Government would be far more extensive. Science advising had become much more complex since the days when the PSAC was founded.[17] When Eisenhower brought his Science Advisor into the White House in 1957, military outlays had commanded the bulk of federal funding for science and technology. But by the 1980s, the primary justification for government involvement in science and technology was their contribution to the national economy, education, health, the environment, as well as national security.[18]

The commission focused on better channeling of scientific knowledge into policy decision making for scientifically illiterate political leaders in various aspects of government. It formed task forces on science as it related to government branches from the courtroom of the Judiciary to international affairs, diplomacy, global development, and environmental decisions and regulation.

The commission ran at a cost of about 13 or 14 million dollars over five years. Hamburg convinced Lederberg to co-chair it with William Golden. As President of the New York Academy of Sciences and Chairman of the Board of the American Museum of Natural History in New York, Golden had also played an important role in establishing the original PSAC in the early 1950s. There were two other Nobel laureates on the Commission along with two former US Presidents: Gerald Ford and Jimmy Carter.

There was plenty of criticism when it issued the results of its study in 1993.[19] Some policy analysts criticized the commission for wanting "a scientific advisor to everyone and his uncle," or that they wanted to turn scientific advisors "into a select priesthood" with "influence beyond political legitimacy."[20] The commission discussed those criticisms in various reports.[21] Other critics pointed to the makeup of the committee — almost entirely of white men over the age of 50. There were only three women on the 22-member commission and two on its 31-member advisory council.[22] Hamburg acknowledged that if he had to do it all over again, he would have included more young scientists and involved more women and minorities.[23] Younger committee members such as Jesse Ausubel, Director of the Human Development Program at the Rockefeller, lamented how the older members reminisced about the Cold War and the Kennedy era and so on, with little appreciation of what younger people might have to offer. Young people would be "the vectors of change," not just the reports.[24]

Still, when the Commission's final report was distributed to state and federal policymakers, it had an immediate impact on the organization of the federal government. In fact, several members of its task force were appointed into the Clinton Administration. For example, aeronautics specialist Sheila Widnall was nominated

Secretary of the Air Force in 1993, the first woman to hold that position. William Perry, who had chaired the commission's task force on national security, became the Deputy Secretary of Defense that year.

There were other problems too that appeared along political party lines: Republicans tended to view scientific advice as unnecessary.[25] In the mid-1990s, the Office for Technology Assessment (OTA) was eliminated. The OTA had helped shape science and technology policy for 24 years, conducting some 750 studies on a wide range of topics, including acid rain, health care, and global climate change. The OTA was weakened during the Reagan administration, and when the Republican-led Congress of Newt Gingrich planned to eliminate it altogether, Lederberg tried to dissuade him.[26] All attempts to save it were defeated in the Senate, however. Its elimination was a major setback for science policy as the Carnegie commissioners saw it.[27]

Science advising got a boost during the Clinton administration. The Carnegie commissioners were gratified when the Secretary of State, Madeline Albright, established an Office of Science and Technology Advisor in the State department. President Clinton also made an historic appointment in nominating nuclear physicist, John Gibbons, as Assistant to the President for Science and Technology, and director of the Office of Science and Technology Policy.

Unfortunately, this new version of the PSAC was underemployed. The problem, as Lederberg saw it, had to do with the very issue that led him to suggest the Carnegie study in the first place: confidentiality. He had hoped that a statement of confidentiality could be written such that the President would have enough trust in the committee to fully employ it. But by 1998, that had not eventuated. "We have PSAC," Lederberg lamented, "but the

President doesn't trust it and doesn't do much with it, and makes many of his appointments on grounds of diversity, as much as on trying to get the expertise that he's looking for."[28]

By that time, Lederberg had advised various agencies within the Federal government — the Pentagon, the State Department, and National Security Council — on the threat of biowarfare, emerging diseases, and the need for public health programs to protect the public against an epidemic, whether accidentally caused or not. He was well respected, admired for his intellect and approach to policy matters, and had the highest national security clearance. He also had first-hand experience with the reluctance of government officials in accepting expert advice. But in the 1980s, his main work for national and international security was liaising with Soviet scientists in an attempt to ensure compliance to the biowarfare treaty he had helped to establish.

Chapter 27

Soviet Secrets

> My bottom line is fairly gloomy. The new technologies that may reopen BW for strategic conflict will come primarily from *medical*, not military research. During the next decade, it is US researchers who will be uncovering the biology of virulence and of host-specificity; and that publicly available work will be capable of fairly prompt breakout by any side that has that intention.
>
> Joshua Lederberg on his meetings with Soviet Scientists, June 1985[1]

The biological weapons and toxins treaty of 1972 relied on the good faith and commitment of the parties who signed it. But it also suggested mutual consultation among experts to help ensure compliance. Lederberg took to that forum over the next two decades consulting with Soviet scientists and with the US government. His first mission though, was in preventing further biowarfare development in the United States.

The biowarfare treaty allowed for research and development for defense purposes only. The United States had abolished its offensive bioweapons program at the time the treaty was signed, but in the late 1970s, there was renewed interest in its potential further development in American government circles. In 1977, when the Undersecretary of Defense, William Perry, was advised that he should be pursuing a biological warfare program, he wanted to get some expert advice first.[2] He needed someone who had been a consultant for the defense department, had security clearances, and therefore could be talked to freely. He asked around and got the name "Josh Lederberg."

"So I called him," he recalled, "and asked him if he would come down and spend a Saturday morning with me, and that is when I did my thinking. For most of the rest of the week I was on a treadmill." They spent half the day in deep discussion on the issue, and Lederberg convinced him that it was a bad idea. Perry was enormously impressed with how Lederberg reasoned — not on moral grounds, but on science and effectiveness; on logical arguments, not on emotion. He put himself in the position of a military commander who would have this weapon, and how he would use it and how ineffective it would be. "It was a *tour de force* really," Perry recalled. "So I just killed the program then and there. He was so persuasive, so effective. That was one of Josh's achievements that people don't talk about — that from 1977 onward the US did not have a biological weapons program either visible or invisible."[3] Perry called on Lederberg repeatedly when he became Secretary of Defence; they became lifelong friends.

Lederberg's main security mission in the 1980s was to establish trust with Soviet scientists to verify that they were adhering to the bioweapons treaty and explore means to strengthen compliance. Two suspicious incidents led American government officials to believe that the Soviets had developed biological weapons in violation of international law. The first was called "yellow rain." In 1981, the United States Secretary of State, Alexander Hague, accused the Soviet Union of producing and supplying highly toxic poisons (mycotoxins) to communist states — Vietnam, Laos, and Cambodia — to use against insurgents.[4]

The story of Soviet germ warfare reached the press, and the Reagan Administration badgered the Soviets about germ warfare throughout most of the 1980s.[5] It was the stuff of a propaganda war. But the State Department's evidence with regard to "yellow rain" was doubtful.[6] It was based on reports of chemical attacks on villagers in Laos who had described clouds of "yellow rain," which they said caused nausea, vomiting, diarrhea, and death.

Lederberg and Matthew Meselson worked together as they had before on the biowarfare treaty in the 1960s.[7] "More urgently than ever, I feel the need to find some way to reestablish mutual trust with the USSR on verification issues," Lederberg wrote to Meselson in June 1982.[8] The whole basis of the "yellow rain" accusation would soon come undone, at least as far they were concerned, as "yellow rain" itself seemed to be due to natural causes.

First, microscopic studies indicated that its principal component was actually pollen from plants that were common in Southeast Asia.[9] Its pollen content suggested that bees were somehow involved, and bee experts that Meselson contacted suggested that the yellow spots were, in fact, bee feces. Still, no one knew whether tropical Asian honeybees defecated collectively, producing showers that could be mistaken as spray from an aircraft. In March 1984, Meselson and his team went to Thailand and observed this very thing.[10]

The case for the second suspicious incident indicating that the Soviets had violated the treaty was much stronger. The Frankfurt-based Russian newspaper *Passev* first reported the story as told by Soviet dissidents in January 1980: an anthrax outbreak had occurred in April and May of the previous year that was caused by an explosion at a military facility in the city of Sverdlovsk, 900 miles east of Moscow. Military medical personnel were brought in to replace civilian workers in the area of the hospital set aside for the victims exposed to the anthrax spores.[11] The Soviet defectors also said that there were other secret biotechnology-based biowarfare research operations in several Soviet cities in addition to Sverdlovsk.[12]

The Soviet government denied all of this and maintained that the anthrax outbreak in Sverdlovsk did not result from airborne spores, but rather from villagers eating meat from infected black market livestock sold in violation of veterinary regulations. The US government rejected that explanation.[13] And despite repeated requests for further information about the anthrax outbreak, by the mid-1980s, none had been supplied.

Lederberg had long feared that genetic engineering might be applied to germ warfare, and not just to make superbugs. Lederberg mentioned some of the other horrors he could imagine to Vint Cerf, one of the founders of the internet, when they was at Stanford in the 1970s. In one scenario that Lederberg presented, a virus was used to infect soldiers. It had the capability of editing their DNA so that they would produce endorphins in unlimited quantities. "Could you imagine," Cerf exclaimed, "what would happen to the opposition's military if they were doped up on opiates on a permanent basis and they couldn't stop it because it was part of their DNA? Holy crap, you know that made quite a scenario for a story."[14]

Lederberg aimed to gain further information on what happened in Sverdlovsk and find ways of strengthening compliance with the treaty with Soviet counterparts. First, he convinced the US National Academy of Sciences, which had a Committee on International Security and Arms Control that focused on nuclear arms, to allow him to form a subcommittee focusing on biological weapons prevention. Then he established meetings with a delegation from the Soviet Academy of Sciences.

In June 1985, he and his group travelled to Moscow to meet with Soviet counterparts on a four-day mission.[15] His discursive strategy for obtaining Soviet cooperation and openness was to gain trust with his counterparts based on a common scientific culture, and to insist that the USSR and the US had a mutual enemy in the threat of bioterrorism.

The meetings were intense. Soviet scientists knew of Lederberg. He was held in esteem by some, and was defamed by others.[16] In an address to the Soviet Academy of Sciences in 1981, physicist Aleksandr Aleksandrov falsely accused him of supporting "race-prejudiced views on human genetics and with suggestions to punish the poor." Nothing could have been further from the truth. Lederberg wrote to him: "I have denounced misconceptions of genetic fatalism on many occasions… Furthermore, I have expressed the greatest skepticism about the importance of genetic factors playing any important role in economic accomplishment by race, religion, class, or any other major cultural grouping."[17] Lederberg received no reply then, but as it turned out, he would have the opportunity to refute Aleksandrov's charges on Soviet television in the spring of 1985.

The main concern of the Soviet scientists was President Ronald Reagan's Strategic Defence Initiative — that is, the anti-ballistic missile system, dubbed "Star Wars." Some Soviet scientists warned Lederberg and his group that Reagan's Strategic Defence Initiative would create instability and that the Soviets would have to escalate their conventional and unconventional warfare abilities and, perhaps, develop biological and chemical weapons.[18]

Lederberg got a mixed response when he showed the US news report about the anthrax outbreak near a military facility in Sverdlovsk indicating that the Soviet Union already had such a program. Aleksandr Baev, who directed recombinant DNA research at an institute on the outskirts of Moscow, dismissed the allegations and insisted that health authorities could give him the correct information.

Other, more internationally minded Soviets, were receptive. Nicolai Bochkov, Director of the Institute for Medical Genetics who was involved with "Doctors for the Prevention of Nuclear War," agreed to pursue his own information about Sverdlovsk.[19] Cambridge-educated physicist Sergey Kapista arranged an interview with Lederberg and Bochkov to discuss human and medical genetics on his UNESCO award-winning television program, *Evident, but Incredible.* Lederberg found that they were in complete agreement on such matters as the genetics of IQ, the prospects of genetic diagnosis for prenatal disease, and the general nature-nurture problem.[20]

Lederberg's most important meeting was with molecular biologist Yuri Ovchinnikov. He was Vice-President of the Soviet Academy of Science, a member of the Central Committee of the Communist Party, and in charge of all biological facilities in the Soviet Union. He knew well of Lederberg, and a few years earlier had asked him for a photo that he could use in a textbook he was writing then.[21] They met at Ovchinnikov's grand and new "Institute of Bioorganic Chemistry" — a superbly equipped facility, constructed with a budget of 300 million dollars and maintained with an annual budget of 100 million rubles. Ovchinnikov told Lederberg that he had gotten Brezhnev's personal backing to modernize Soviet biology through molecular genetics because of "its indispensible value for medicine and agriculture."[22]

Lederberg was pretty sure that biowarfare research was being carried out and he was on the lookout for signs at Ovchinnikov's institute.[23] He had been given word that they were to "talk strictly science." But the conversation veered to biological weapons when they all went to lunch, and one of Ovchinnikov's colleagues asked Lederberg about his Committee for International Security and Arms Control.

Lederberg explained that biological warfare was perilous and also self-destructive. What was needed, he said, was a coordinated effort with the Soviet Union to manage the biological warfare treaty on a global perspective. He pointed out, for example, that terrorists could easily contaminate a city's water supply with home-grown typhoid. Ovchinnikov agreed and also spoke very articulately about the hazards of biological warfare proliferation and that "every step should be taken to prevent microbes from being used as weapons."[24]

Ovchinnikov's colleagues seemed to know nothing of the anthrax outbreak at Sverdlovsk. That news was kept from them. Nothing about it was published in the USSR — even *Science* magazine was censored to all but a few senior academicians, such as Ovchinnikov, who would certainly know what had happened at Sverdlovsk. He didn't respond when Lederberg urged him to get better information on it. Instead, he pressed for other kinds of scientific cooperation with American molecular biologists. Lederberg also wanted that cooperation as a way to gain trust, but that would be impossible as long as the Soviets refused to give a proper account of what happened at Sverdlovsk. Until then, the US State Department would continue considering it as evidence of a treaty violation. The Soviets had broken international law, plain and simple.

Lederberg concluded his report of his Moscow meetings with the depressing statement that the research fostering biological warfare was going to come, not from military research, but mainly from medical research on virulence and on host-specificity and that publicly available research could be used by any side for biowarfare. "Secretive work certainly does speak to hostile (or defensive) intentions; and we would be better off were we able somehow to get to more open communication about work in the microbiological area," he wrote. "I believe this visit did communicate that message."[25]

Those first meetings were steps in the right direction in so much as they put biological weapons control and the enforcement of the biological weapons treaty on the agenda.[26] Just having such discussions on biological warfare with top-level Soviet scientists was an impressive feat as far as the State Department and the National Academy of Sciences were concerned.[27] Lederberg continued to work in dialogue with Soviet counterparts over the next several years.

In the fall of 1986, Lederberg and his team made a second visit to Moscow.[28] Those meetings were held a few days before the famous encounter between Mikhail Gorbachev and Ronald Reagan on nuclear arms control. Gorbachev proposed banning ballistic missiles and Reagan wanted continued research on the Strategic Defence Initiative. Though no agreement was reached, it marked a turning point in the Cold War. Lederberg's meetings with Soviet scientists were no more or less successful.

They met at Ovchinnikov's Institute in Moscow. Lederberg especially wanted his cooperation but, unfortunately, he was said to be out of the country.[29] Ovchinnikov's absence was not encouraging that things would go well. Then things seemed to take another step backward when the Soviet representatives at the meeting pointed to reports in the American press the month before that the US Defense Department had increased spending on biological weapons development and testing by 40 million dollars.

This time, Lederberg was on the defensive. He explained that there was no offensive biological weapons program in the US and that the facilities at Dugway were under intense debate and close scrutiny by Congress. (Funds could be used for defensive purposes under the treaty, such as to make vaccines to inoculate solders against an attack of anthrax or botulinum.) He countered Soviet scientists' concerns about that by arguing that the very existence of that public information in the US contrasted with the absence of such public information about Soviet activities. The lack of openness only led to "speculation with a tendency toward worst case scenarios" that fed a technology race.[30]

The Soviets did, however, offer a more detailed account of the anthrax outbreak at Sverdlovsk at that meeting. A physician from the Soviet Union's Ministry of Health showed up to give a two-hour lecture showing autopsy slides of the victims. Following the Soviet story Lederberg had heard on his previous visit, he maintained that there was no evidence of airborne anthrax and that the intestinal anthrax was from infected meat sold on the black market.

Lederberg did not know how much it could be trusted. But even if it were true, as he informed the US Army when he returned from Moscow in November 1986, there were "ample other reasons to believe that the Soviets continue to sustain a substantial BW development effort."[31] A few months later, he learned from anthrax experts in the United States that, in fact, some of the symptoms the Soviets described were not those of intestinal anthrax, but actually of inhalation of airborne anthrax.[32]

A third meeting with Soviet scientists was held in the fall of 1987, but this time at the Rockefeller University.[33] Lederberg's aim was to allow his Soviet counterparts to actually participate in assurance of compliance in the United States. He knew the Undersecretary of Defense, Frederick Iké, at the Pentagon, having worked with him as an advisor on Soviet relations and defense strategies.[34] And he managed to get his permission to arrange a visit to Fort Detrick in the hope of setting up a reciprocal exchange of visits — including to the Soviet military facility at Sverdlovsk.[35]

In the meantime, Lederberg's quest for transparency was threatened by disinformation in the press. The KGB had a huge program of spreading false stories to confuse citizens of Western democracies, just as it does today, and it planted a story in the world press that AIDS was a lethal weapon of the Department of Defence that was developed at Fort Detrick and had been designed to kill African Americans and homosexuals.[36] This time, Lederberg's Soviet counterparts criticized the Soviet media for promulgated false claims.[37]

Things got potentially bad again, as Lederberg saw it, when in October 1989, some 800 American scientists signed an ill-informed pledge circulated by the Council for Responsible Genetics "against the military use of biological research" in the United States. Nobel laureate biochemist Christian Anfinsen at Johns Hopkins University wrote to Lederberg asking for his signature. "Because the biological weapons convention does not now include verification measures," Anfinsen wrote:

> advances in our fields have led to increased mistrust between nations and suspicion that biotechnology may be exploited for offensive purposes under the guise of defensive research. This mistrust can be seen in the policies of the last administration in the US which quadrupled funds for research in the Army's biological defense program. And suspicion can be seen in stepped-up research efforts in other countries as well.[38]

Lederberg was dismayed by Anfinsen's petition.[39] He knew Anfinsen well enough; they had corresponded about phage research since the late 1950s. But Anfinsen seemed unaware of how much Lederberg had become involved in the formation of the biowarfare treaty of 1972, and of his committee which aimed at gaining trust and compliance. Of course, as he explained to Anfinsen, he was sympathetic to all "the well meaning people" who signed the pledge. Biological weapons development was a "very grave risk in an unstable world and one which deserves all the ethical opprobrium that lies behind your solicitations."[40] But, the pledge, as it was worded, was simply wrong-headed and would do more harm than good, as he saw it.

The most troublesome part was the petition's pledge against the "military use of biological research." Surely that was not what they were protesting? After all, there were military uses of biological research in a range of contexts including medical care for US troops, especially against infectious disease, which Lederberg insisted constituted "the overwhelming part if not the totality of the Army's biological research programs." "I wish you had taken more time to inform yourself about the details of biological research in the military, before putting the great weight of your name behind the document," he replied. The assumption that the increase in research funds by the Reagan administration was aimed at escalating biowarfare development was wholly unfounded. He urged Anfinsen to visit Fort Detrick to learn about the research programs there.[41]

The timing of the petition was also troublesome for Lederberg. He was worried that attacking the army in the midst of glasnost would be detrimental to his own negotiations. "We are right in the middle of extensive discussions with Soviet biomedical scientists," he wrote to Anfinsen,

> trying to get them to understand the need for as much public disclosure of their biological research programs as exists in the US. In the wonderful new spirit of glasnost, great progress has been made in this

> direction. They are just beginning to open up. I am concerned that the political reactionaries in the Soviet Union will feel that they can make hay out of our domestic attacks against the US Army's program and slow up the process of glasnost with respect to their own activities.[42]

When Anfinsen met with Lederberg in New York two weeks later, he acknowledged that he was not well-informed.[43]

Lederberg saw the promotion of openness through the exchange of scientists as the most useful confidence-building measure. In April 1989, his committee met with Soviet scientists to refine what kind of research was permitted or banned under the bioweapons treaty.[44] The shock came just six months later when the story emerged that Leonid Brezhnev had indeed set up an enormous biowarfare program. In October 1989, the top Soviet biologist Vladimir Pasechnik, who had defected to Britain, told interrogators that thousands of scientists worked on biological weapons under false fronts across the Soviet Union. Their program consisted of everything Lederberg had feared: genetically engineered pathogens. They had made a superbug that was genetically engineered to be resistant to heat, cold, and antibiotics, and he also revealed that long-range weapons had been created to deliver it.

Pasechnik's testimony was soon confirmed. Gorbachev had been in office for a year before he learned of the scope of the program; they had even tried to keep it from *him*.[45] Ironically, it turned out that the Soviet program had been induced by Nixon's announcement at the end of 1969 to terminate the United States' offensive biological weapons program. The Soviets did not believe a word of it, so they did the opposite: they escalated theirs.[46]

Yuri Ovchinnikov, whom Lederberg had met in his first mission to Moscow in 1985, had been the head of it all.[47] It was called

Biopreprat and was established as the "civilian" continuation of earlier Soviet bioweapons programs.[48] Ovchinnikov promoted the military benefits of genetically engineering bioweapons as a strategy to fund his own molecular biology program aimed at medical products such as interferon for cancer and biosynthetic insulin, and he got support from the Soviet government by arguing that the US was likely to apply new genetic engineering methods to create deadly pathogens as weapons.

The US intelligence account of pulmonary anthrax radiating from a bioweapons facility was confirmed in 1994 after Meselson and colleagues were finally permitted to travel to Sverdlovsk, where they were able to track the outbreak. The anthrax spores had been accidentally released from the military microbiology facility due to botched air filter maintenance.[49] It was one of the largest inhalation anthrax outbreaks in history, the biowarfare equivalent of Chernobyl. Sixty people died along with many livestock. Had the winds been blowing toward the center of town that day, thousands of people could have been killed. Regular anthrax can be treated with antibiotics and prevented with vaccines, but the Soviets were working with an antibiotic-resistant strain that could also evade vaccines.

President Boris Yeltsin called for an end to the offensive biological warfare program in 1992, a few months after the Soviet Union dissolved. When Vladimir Putin became President in 2000, Russian officials lied, stating that it was strictly a defensive program and thus had not broken international law. There is suspicion that he renewed the biowarfare program.[50]

In October 1989, President George H. Bush awarded Lederberg the National Medal of Science with the inscription: "For his work in bacteria genetics and immune cell single type antibody production; for his seminal research in artificial intelligence in biochemistry and medicine; and for his extensive advisory role in government, industry and international organizations that address themselves to the societal role of science."

Lederberg turned to work on a new risk that emerged from the fall of the Soviet Union. Russian scientists involved in military research might be enticed by third parties and sell themselves to the highest bidder.[51] As one of the few biologists with any standing among Pentagon officials, Lederberg was able to overcome Pentagon resistance against funding joint US-Russian research projects to reorient Russian bioweapons work for peaceful purposes.[52] When he and his group met with a Russian Academy delegation in 1993, he said it was time for cooperative work rather than struggling to develop a trickle of information between hostile powers.[53] Their new aims were to discuss plans to destroy smallpox strains and repurpose military-related facilities for civilian uses.

Fig. 27.1. Committee on International Security and Arms Control Moscow 1991. The Joshua Lederberg Papers.

Fig. 27.2. President George H. W. Bush awards Joshua Lederberg The National Medal of Science, October 18, 1989. The Joshua Lederberg Papers.

At the same time, Lederberg was heavily engaged in raising the alarm about microbial dangers and civilian vulnerabilities to outbreaks at home. The main threat from biowarfare, in Lederberg's view, was from terrorist groups. Several of them had already used bioweapons. However, bioweapons were only part of the threat. Humans, he warned, were also beginning to lose the battle against infectious disease.

Chapter 28

The Top Predator

> The human occupation of planet Earth has left us with a conceit that our species is at the very top of the food chain, predatory on all others, rapidly destroying their habitats… Barring genosuicide, the human dominion is challenged only by the pathogenic microbes, the predator for whom we remain the prey.
>
> Joshua Lederberg, 1993[1]

Greek mythology tells the story of the unrequited love of a heart-broken demi-god who wanted to punish the princess Cassandra for rejecting him. He bestowed on her the gift of prophecy and then he cursed her never to be believed. Lederberg sounded the alarm about coming plagues beginning in the 1960s with the then newly reported Marburg virus and Lassa Fever. He connected those outbreaks with his own fear that pathogens might be developed as biowarfare agents. Epidemics and biowarfare were joined at the hip — whether accidental or deliberate, they were threats to the nation, and the world, and required research and public policy to prevent.

Lederberg's first warning was made in August 1968 in his *Washington Post* article, "The Infamous Black Death May Return to Haunt Us."[2] Then, the next week, he explained that "Mankind had a near miss from a Mystery Pandemic" — caused by the Marburg virus — and that it would not be long before "encounters with such threats would not be just near misses." He advised readers that science was ill equipped to stop a global pandemic.[3] He addressed the threat again in *The New York Times* in January 1970, arguing that molecular studies of viruses would be critical in preventing another great pandemic, like the bubonic plague of the 14th century which wiped out one third of Europeans, or the flu pandemic of 1918–1919 which killed 25 million.[4]

Unfortunately, Lederberg's warning of "natural" plagues fell on deaf ears. To most readers, the war against infectious diseases was already won through vaccines, antibiotics, sanitation, health education, and better nutrition.[5] The warning of plagues to come seemed to be ludicrous and outdated. A few decades earlier there had been lobar pneumonia, meningococcal meningitis, streptococcal infections, diphtheria, endocarditis, enteric fevers, various septicemias, syphilis, and tuberculosis everywhere. But these afflictions now seemed like nightmares from the past. Lewis Thomas, the president of New York's Sloan-Kettering Cancer Center, told his readers in 1974 that such fears were "paranoid delusions on a societal scale, explainable in part by our need for enemies, and in part by our memory of what things used to be like."[6]

That sense of having conquered microbial infections seemed to be shattered with the arrival of the HIV pandemic in the 1980s. Lederberg warned his readers again that its emergence was not a unique, one-off phenomenon. Some infectious diseases, he predicted, would newly appear; others would increase in incidence and expand in geographic range.[7] Declarations that infectious diseases had been conquered stemmed from a false, non-evolutionary framework that ignored or misunderstood the natural history of infectious diseases. The public health community, according to him, effectively viewed disease from the perspective of special creation.[8]

The reception to Lederberg's warning about emerging disease began to change, at least in the world of academe in 1987, when he gave a seminal talk titled "Pandemic as a Natural Evolutionary

Phenomenon" at a conference on the social and ethical issues of AIDS at the New School for Social Research in New York. "There was no guarantee that we would always find ourselves the winner in the evolutionary competition of viruses with the human species," he said.[9] His views were partly derived from what he had observed in the laboratory. There, he witnessed a population of a billion bacteria, living in an uneasy equilibrium with their viral parasites, being suddenly wiped out and replaced by 100 billion viruses. That was a microcosm of the human species living under "a kind of" Sword of Damocles in a test tube. The fundamental principles would be the same.[10]

In the long run, most microbes have a shared interest in their hosts' survival. After all, killing a host is a pyrrhic victory for most microbes because a dead host is a dead end. Evolution thus leads to a co-adaptation of less virulence of the pathogen on the one hand and increased host resistance on the other. But viruses can mutate and evolve at a rate that is incomparable to the slow evolution of their animal hosts. We can't be sure that the influenza pandemic of 1918 was not due to an extraordinarily virulent virus, unlike any since, that might appear again.

Lederberg warned that the threat of pandemics was not solely a matter of new pathogens resulting from mutations though. Overpopulation and the opening of wild lands to human occupation can lead to the transfer of viruses from animal hosts where they had co-adapted and symptoms are mild to human populations where they are lethal. Ebola and yellow fever were sustained in jungle primates, and primates were probably also the vectors of HIV in Africa. The "opening" of the Amazon basin, Lederberg prophesied, was almost certain to lead to new disease outbreaks.

The case of HIV, which attacks the immune system itself and encourages other infections, had the world on edge for fear of what might come next. Would the virus spread further? What were the prospects of a vaccine? "I was labeled an alarmist twenty years ago for raising a 'specter' of pandemic," Lederberg commented in 1988. "My most pessimistic imagination did not fetch the constellation of attributes that we observe with AIDS."[11] There was nothing but uncertainty about AIDS in 1988. The virus was still evolving and the existence of various strains further complicated the task of developing a vaccine. But Lederberg raised the possibility of an even greater threat. What if the virus changed its mode of transmission and became airborne?[12] The ramifications were unthinkable.

Lederberg suspected that AIDS had been spreading in Africa for 10 or 15 years unnoticed.[13] The pandemic was due to the neglect of infectious diseases in the poor majority of the world where tuberculosis and malaria were still deadly rivals of AIDS. Such carelessness was not only a humanitarian disgrace but also seeded diseases that affected the developed world.[14] There needed to be a systematic watch for other new viruses before they led to pandemics.[15]

Lederberg's presentation at the New School was well received.[16] So he decided to give another presentation on emerging infectious diseases a few months later in January 1988, this time to an extraordinary audience of the world's Nobel laureates who gathered at a conference held at the Élysée Palace in Paris. President François Mitterrand and Elie Wiesel, Professor of Humanities at Boston University, were its hosts.

Wiesel had been awarded the Nobel Peace Prize the year before for speaking out against violence, repression, and racism. The Nobel committee called him "a messenger to mankind." He had campaigned for victims of oppression in South Africa, Nicaragua, Kosovo, and Sudan. He had also authored more than

50 books; his most famous, *La Nuit*, was an autobiographical account of his horrific experiences with his father in the Auschwitz and Buchenwald concentration camps. Written in 1955, *La Nuit* (*Night* in English) is considered a masterpiece at the core of Holocaust literature.[17] In 1978, he chaired Jimmy Carter's Commission on the holocaust, which led to the creation of the Holocaust Museum in Washington.

Lederberg had not yet met Wiesel when he received his invitation to the Paris meeting. But he knew him as "the poet of the holocaust" and had, in fact, nominated him for the Nobel Prize in literature in 1973.[18] "The real danger," Lederberg wrote to the Nobel Committee, "might have been that the holocaust remains unforgettable but totally beyond human comprehension and to that extent put out of mind as a matter outside the human experience. There are grave tendencies through the years and today that continue to encourage such a repression, in effect such a dehumanization of history."[19]

Despite Lederberg's esteem for Wiesel, he was not optimistic about the aims of the Nobel conference in Paris, which, as Wiesel wrote to him in July 1987, was about "the principal moral and political problems that challenge humankind."[20] He did not share Wiesel's views about Nobel Laureates being such wise and humane people, or the value of collecting them "to make some statements about how the world might be made better through good thinking," as he put it a few years later.[21] But he thought it was important to support Wiesel as "a symbol of world repentance over the holocaust," and he thought his acceptance might help to "encourage Mitterrand in his stance against the French anti-Semitism which was starting to become ugly."[22] Wiesel asked Lederberg to give one of the plenary presentations.

In preparation for the event, Lederberg read *Einstein on Peace,* a collection of Einstein's writings on war, peace, and the atomic bomb, his views as a socialist, and his internationalism and opposition to nationalism, which he saw to be responsible for many evils of the world.[23] Lederberg found *Einstein on Peace* to be "deeply inspirational" but also "disturbing in many ways."[24] The troubling part was that Einstein's deepest convictions were repeatedly overturned by history. He had been a pacifist from 1914 until 1933; then, in 1939, he sent the famous letter to FDR, urging him to begin building an atomic bomb before Nazi Germany did. After the war, he spent much of his time trying to undo the consequences of those nuclear weapons.

Lederberg vehemently opposed communism, and he suspected that Einstein was also "not totally taken in by the myth of communism." But like Einstein, Lederberg was also committed to internationalism, though a world government seemed quixotic: Einstein "was absolutely right in fixing on the central problem of trying to find a successor to national sovereignties and of course achieved absolutely no success in that direction."[25]

It was a grand and glitzy affair in Paris in January 1988. Josh and Marguerite were attended by the Republican Guard and by motorcycle escorts through Paris. They stayed at the lavish Hôtel de Marigny, reserved normally for heads of state, not far from the Élysée Palace where the conference was held. The pomp and ceremony were perfect, but the quality of the talks and seminars was mixed, just as Lederberg had suspected it might be.

The plenary meetings occupied a couple of hours a day. Another four hours or so were spent on the smaller, specialized sessions, most of which he thought were awful.[26] Among the most irritating was that of Brian Josephson, who, after being awarded the Nobel Prize in physics in 1973, took up transcendental meditation, explored the idea of intelligence in nature, and aimed to synthesize science and eastern mysticism. Lederberg thought Josephson just "made a fool of himself… with his gibberish about laying on of hands for healing."[27]

The pithiest discussions at the conference focused on "Third World debt and its remission," which Mitterrand had called for. Lederberg found other discussions about disarmament and peace to be "pretty shallow." He did think that Henry Kissinger's talk at the Paris meeting was remarkably eloquent though.[28] He had no doubt that he himself had also communicated very effectively about the natural history of pandemics and the AIDS epidemic. Several members of the audience told him how they were moved by his speech.[29]

He cautioned that the AIDS epidemic was not a one-off episode. It was "a natural, almost predictable, phenomenon" that would recur, so international cooperation and a systematic watch for new viruses were needed.[30] Viral research, especially using recombinant DNA methods, which, he said, was "still a scare word in some quarters, is our most potent means of analyzing viruses and developing vaccines":

> As one species, we share a common vulnerability to these scourges. No matter how selfish our motives, we can no longer be indifferent to the suffering of others. The microbe that felled one child in a distant continent yesterday can reach yours today and seed a global pandemic tomorrow. "Never send to know for whom the bell tolls; it tolls for thee."[31]

There were also some good personal by-products of the meeting. Marguerite was seated next to Mitterrand at one of the dinners and was able to tell him about her girlhood experience hiding from the Nazis in France. She explained that her father had been Mitterrand's father's personal physician, and that Mitterrand's sister,

Antoinette, had been most helpful to her mother during the war. Lederberg also had a good talk with Kissinger, learning his views on the capabilities for discriminate nuclear strikes to defend a limited nuclear attack, the Strategic Arms Limitation Treaty, and geopolitics.

Lederberg and Wiesel also became close friends and mutual admirers. Lederberg's talents awed Wiesel. Three months after the Paris meeting, Wiesel wrote to him: "Your essay is superb. The relationship between science and religion has rarely seen explored in such depth. Not only are you a great, a very great scientist, you are also a gifted, a very gifted writer."[32] It's hard to know what essay he was referring to.

Few scientists had written on more diverse topics than Lederberg, but he had not written much on religion. The only exception I know of was his piece in *The Washington Post* in 1970, in which he suggested that ecology had the virtues of being an important religion that could replace the anthropomorphism of Judeo-Christian theology and the biblical lore of human dominion over nature: "Is it 'God's will that man exploit nature for the proper ends,' as we are taught by the Bible, the US Constitution and the dialectical materialists alike? Or does our scientific knowledge of man as a stage in the evolutionary process lead us to a new appreciation of man's place not over, but in nature?"[33]

While others had derided the ecology movement as having the flavor of a religious revival, Lederberg celebrated ecology as having "all the requisites of an authentic religion." "We live today in a vacuum of faith," he said. "The love of earth can be at once the most primitive and the most sophisticated of religions, and it deserves the same respect as the other credos by which men shape their lives. As with other religions, its slogans may also require creative reinterpretation before they are either criticized or routinely applied to daily life."

Lederberg argued that whereas in the time of "St Francis Assisi, this theology responded primarily to spiritual needs," today it is one of the most materialistic of concerns about how humans can survive to enjoy their own creations. "The image of nature is deeply rooted in every man's axioms of beauty... While one could disagree about the merits of one style of painting or architecture, no one we deem sane is likely to deny the beauty of the unspoiled landscape." Perhaps this was the essay Wiesel was referring to.

The AIDS pandemic gave Lederberg a stronger voice in his warnings about plagues to come. He also went into action a few weeks after Paris, and by the next spring, the Rockefeller University and the National Institutes of Health had sponsored the first conference on emerging viruses. That meeting was conceived at a cocktail party that was held for Rockefeller University faculty at the president's house. Lederberg met a junior faculty member there, Stephen Morse, who was also interested in emerging viruses.[34]

Marguerite played matchmaker. As Morse recalled, she said to Lederberg, "Sweetheart, didn't you have some questions about virology? Steve's a virologist, you know." Josh went "Oh yes" and proceeded to ask him whether there was any concern about Hantavirus exposure for those working in the university's animal facilities, a problem that Carleton Gajdusek had mentioned to him over dinner at the meeting in Paris.

Gajdusek was interested in hemorrhagic viruses, such as Marburg, Ebola, Lassa, and yellow fever, as well as Hantavirus, which was first isolated in 1976 after an outbreak in the Hantan River area in South Korea. The virus is contracted by inhaling aerosolized rodent excreta (urine and feces) that had been contaminated by the virus. There were school children in Russia who had contracted hantavirus from laboratory rats while touring animal facilities on a school trip.[35]

Gujdusek and his team had discovered the first American form of the virus in wild rodents in 1982. Upon Lederberg's prompting, Morse looked into it at the Rockefeller and was relieved to learn that there wasn't a problem, and that all their rodents were routinely tested. Lederberg wrote a memo back to Morse: "I am of course reassured," he said, and further added, "We need some high-level policy attention to what needs to be done globally to deal with the threat of emerging viruses, and I would welcome your thoughts on that."[36] Lederberg told Morse that he would back him very strongly if he wanted to organize a conference on the issue.[37] Morse wasted no time in getting people at the National Institutes of Health interested.[38]

The conference was a grand affair in May 1989 at the historic Hotel Washington located near the White House. About 150 people gathered in the large ballroom, including several Nobel laureates. Lederberg opened the three-day meeting with a keynote address. "Nature isn't benign," he told his audience. "The survival of the human species is not a preordained evolutionary program."[39] The main question about emerging viruses was basic: "Is there a scientific basis for concern?" The answer was a resounding "Yes." The number of Americans dying of infectious diseases was rising, antibiotic-resistant diseases were spreading, and the AIDS epidemic was raging.

Lederberg got into an adrenaline-inducing argument when he noted that the evidence that HIV could not be transmitted person-to-person by aerosol was rather flimsy, and that more research on it was required.[40] Nobel Prize-winning virologist Howard Temin spoke up and said, "Well, maybe there's a problem but we oughtn't put it this way. We are going to get very inhumane treatment of people who are carrying HIV if they are pictured as being threats to the rest of the community, so let's lay that down entirely."[41] Lederberg was very sympathetic to what Temin had to say; he wanted effective action, not panic, any more than he wanted panic about biowarfare.[42] Later somebody asked Lederberg, "When should we declare that a newly recognized virus is a new species?" He replied, "When it matters." Morse quoted this to his wife, who was duly impressed and said, "What a Solomonic answer!"[43]

There was deep concern at the meeting about the complacency of the scientific and medical communities, the public, and the political leadership of the United States regarding the danger of emerging infectious diseases, as well as the overall lack of preparedness. The meeting at Hotel Washington had an important effect in getting the National Academy of Sciences' Institute of Medicine involved. In February 1991, it organized a 19-member committee to conduct an 18-month study aimed at identifying significant emerging infectious diseases, determining how to deal with them, and recommending how future outbreaks might be confronted to lessen their impact on public health.

That committee was co-chaired by Lederberg and distinguished virologist Robert Shope. They saw eye-to-eye on much, except on the effects of climate change on infectious disease. Shope briefed President Bill Clinton and Vice-President Al Gore, warning that the range of mosquitoes and other arthropod vectors would increase, affecting the prevalence of dengue, malaria, and other infectious diseases.[44] While Lederberg agreed that there would be some effect there, in his view, they would really be second order and thus neither high on the list of things to worry about nor indeed why we should worry about climate control.[45]

To Lederberg, the effects of climate change on emerging infectious diseases could not even begin to compare with the sheer expansion of our species, high population densities, and worse, populations stratified by standards of economics, nutrition, housing, and public health. In addition, there was an unprecedented mixing of people: a million passengers a day cross national boundaries by air, not to mention the movements of armies, refugees, and mass transportation. "One could hardly have concocted a better-calculated recipe for a tinderbox," he said, "as AIDS already harshly teaches."[46]

As Lederberg saw it, without action taken, the extinction of the species was a possibility. He knew of course that natural selection, in the long run, favors host resistance on the one hand and less virulence on the parasite's part. But that said nothing of what might happen when a virus is transferred from one species to another. What might happen to humans would be much like what happened to rabbits in Australia when the myxoma virus was introduced — a mass die-off.

The myxoma virus lives in equilibrium with South American rabbits; it causes only a mild disease in them. When it was intentionally introduced to control European rabbits in Australia in 1950, it killed 99% of the infected population. This provided a grim illustration of what can occur when a virus jumps from a species that is adapted to a new host that is not. Such an occurrence in humans — a novel virus to which we've never been exposed — is what Lederberg feared most: "I would also question whether human society could survive left on the beach with only a few percent of survivors. Could they function at any level of culture higher than that of the rabbits? And, if reduced to that, would we compete very well with kangaroos?"[47]

The National Institutes of Medicine study that Lederberg and Shope chaired resulted in a ground-breaking book, *Emerging Infections: Microbial Threats to Health in the United States* (1992).[48] It called for improved infectious disease surveillance, better understanding of pathogenesis, and the political will to deal with emerging infections. The book gained some traction in academe. Morse subsequently edited *Emerging Viruses*, and the journal *Emerging Infectious Diseases* was founded in 1994. Morse and colleagues also founded an Internet reporting system to promote global surveillance of infectious diseases.

Lederberg knew that journalists played a key role in getting the message to the public, thereby pressuring politicians to act. In 1992, journalist Richard Preston approached him asking for an infectious disease story that would draw readers. Lederberg told him about an outbreak of the deadly Ebola virus among monkeys at an animal supply company in suburban Washington. Ebola infects primates (including humans) and was first discovered in two simultaneous outbreaks in 1976, one in South Sudan and the other in the Democratic Republic of Congo. When Preston asked Lederberg how bad the outbreak was in Washington, Lederberg explained that army officers dressed in space suits had every single monkey that might harbour the virus killed.[49] Preston published an article in the *New Yorker* where he was a regular contributor and expanded it into his best-selling nonfiction thriller, *The Hot Zone* (1994). The book pushed Lederberg's concerns about emerging diseases into public consciousness as it described the panic of health officials when they realized what the Ebola virus could do if it spread into DC.

In 1996, the *Journal of the American Medical Association* and 35 other journals worldwide planned to document the occurrence, causes, and consequences of emerging and reemerging infections. This resurgence of scientific interest was also matched by the popular media, in magazines, newspapers, best-selling books, TV shows, and movies. Still, Lederberg lamented, there was little response by governments to prepare for an epidemic — whether it was caused unintentionally or by bioterrorism. He continued working hard to change that.

Chapter 29 Restless Farewell

I'd hoped to "retire" from DSB [the Defense Science Board] etc. but keeping nuclear and BW weapons out of the hands of the crazies is too urgent, so I'm still on the Washington National Security Circuit.

Joshua Lederberg to William O. Baker, 1991[1]

Lederberg stepped down as president of the Rockefeller on July 1, 1990. It might well have been a relief to him, but he certainly had not intended to resign that year. He was 65 years old and would have continued on, but after 12 years, Rockefeller faculty wanted a change.[2] In particular, the university needed a president who would be more actively engaged in fundraising. Hiring new faculty and making changes meant dipping into endowment funds, and the university was running a deficit.[3]

Lederberg also seemed to be only a part-time president who spent too much time away, especially in Washington. Sometimes he would take the shuttle between New York and Washington three times a week to offer scientific advice on issues of national security. Very few Rockefeller faculty members knew much about his government activities and many felt neglected.

Lederberg's replacement, Nobelist David Baltimore, brought turmoil to the university almost immediately. He was a wunderkind of immunology and molecular biology, and was ready to bring new luster to the university. Unfortunately, he was forced to resign as president a year later because of an ugly controversy over a publication he had coauthored five years earlier. His collaborator was falsely accused of faking data on the genetic regulation of the immune system.[4] Nobel Prize-winning neurobiologist Torsten Wiesel, whom Lederberg had hired in 1983, was then appointed president. His tenure as president was remarkably successful.

Lederberg remained at the Rockefeller as a professor and established a new laboratory of "Molecular Genetics and Informatics." Led by one long term, researcher, David Thaler, its investigations were diverse. During its first year, its focus was on ways in which the environment inside the cell can influence genetic mutations, and on using computers to help plan experiments based on a DENDRAL-style hypothesis generator he had created at Stanford (Chapter18).[5] The last publication was on "quorum sensing," or how bacteria assess their population density and limit their growth accordingly.[6]

A stream of visiting researchers worked in the lab for various periods of time. Billionaire Larry Ellison, cofounder and CEO of Oracle Corporation, was the most unusual one. Ellison had heard Lederberg speak in 1990 at a Stanford conference focusing on the Human Genome Project, which was formally launched that year. Lederberg's talk was on the complexity of the genome and how computers could aid in analyzing the tsunami of data to come. Ellison was entranced by Lederberg's erudition. "The crystalline clarity and lyrical eloquence of his speech were mesmerizing," he said, "I had never experienced anything like it. I wasn't alone."[7]

They met again two years later when Ellison invited him to his home set in an elaborate Japanese garden on Isabella Avenue in Atherton, California. After a couple of meetings, Ellison gave Lederberg a key to his house, saying he never wanted to hear about him staying anywhere else.[8] They talked of various things including how Ellison's great wealth could be used. Ellison did not seem to be interested in biotech companies, but he was interested in molecular biology, and on one occasion, as *Vanity Fair* reported, Lederberg said, "'Look, if you're so interested in molecular biology, you ought to spend some time in the lab and get your hands dirty.' He said, 'You're on.'"[9] He visited Lederberg's lab for a few weeks in 1994, participating in experiments on bacteria.

Four years later, Ellison established the Ellison Medical Foundation with Lederberg as its chair. At first, the Foundation focused on the biology of ageing. It was not too fashionable an area of research at the time, but with people living longer and being increasingly susceptible to diseases such as Alzheimer's, osteoporosis, and Parkinson's, there was much research to be done.[10] Ellison committed $250 million with the promise that funding would expand to meet demands. "If Lederberg believes that there are projects that should be funded, they will be," he told his biographer.[11] The Foundation expanded to include infectious diseases in the developing world, making treatments for HIV, tuberculosis, and malaria affordable, which Lederberg had long called for.[12]

In Washington during the 1990s, Lederberg turned to domestic defense and headed task forces for the Defense Science Board, which advised the Secretary of Defense. Ten years earlier, he had chaired a task force for the Defence Board aimed at identifying supercomputer and artificial intelligence applications.[13] This time, his task force focused on detection and civic preparedness for a terrorist attack using germ warfare.[14]

No federal agency had done any serious planning for defending against a biological weapons attack on US cities. Lederberg recommended that the Defence Department and the Center for Disease Control and Prevention take the initiative to formulate a plan together.[15] After all, emerging diseases and biological weapons were inseparable in terms of national security; both required monitoring and response procedures to be put in place in case of an outbreak.

Lederberg had long warned of possible terrorist attacks using bioweapons. His warnings caught the attention of the Defence Department when, in February 1993, terrorists attacked the World Trade Center using a truck bomb. Then, in June the following year, the cult movement Aum Shinrikyo used the neurotoxin sarin gas in Matsumoto, Japan, and again nine months later in Tokyo's subway, killing 12 people.[16] New York City was braced for a similar kind of attack with toxins or bioweapons. "The one that scares me to death, perhaps even more so than tactical nuclear weapons, and the one we have the least capability against is biological weapons," General Colin Powell stated at the time.[17] President Bill Clinton agreed.[18] Still, no plan to protect the public was established.

In May 1994, Lederberg became involved in putting together just such a plan in New York City, working with the city's Health Commissioner Margaret (Peggy) Hamburg and Mayor Rudy Giuliani. Peggy is the daughter of the Lederbergs' good friends David and Beatrix Hamburg, and she had known Lederberg since his Stanford days when, as a child, she sometimes brought cookies for their meetings.[19] She remembered going with her mother in the 1960s to see the large computer that Lederberg had in his house, and thinking that it "was like a religious outing" at the beginning of an era.

Hamburg knew Lederberg to be a maverick; she knew his warning about emerging diseases and the need for defense against bioterrorism. "Many people thought that he was an alarmist," she recalled. "Certainly I experienced that too as I got more involved with him and more involved in that issue."[20] She had not taken domestic terrorism seriously until the first attack on the World Trade Towers. Then she began to think about what a germ warfare attack would mean in terms of public health preparedness: "And then you talk to Josh with a vivid imagination and there are so many things…" She began to worry, for example, about the water towers on top of all the apartment buildings. They turned out to be a problem for legionnaires disease, but not for deliberate terrorism.

Hamburg decided to bring Lederberg in to meet with Giuliani when tuberculosis and drug resistance were big problems in New York City. Her office had already embarked on a very successful plan to address these health crises, but she wanted Giuliani "to step up and take credit for it, because she thought it would make him become more engaged in public health as a form of public safety."[21]

Giuliani was concerned with public safety measures involving the police and the fire department, but he was not interested in discussing emergent diseases because he did not want New York to be perceived of as a dirty, disease-ridden city. However, bioterrorism was also a public health issue and it would tie in with Giuliani's interests in law enforcement, so she arranged for Lederberg to meet with him. But then Hamburg began to panic because Lederberg started to give Giuliani a lecture about the life history of anthrax and how it could be spread. As she recalled it:

> Giuliani took us into his about office, and Josh in his way, which could be rather ponderous, started going into the whole life cycle of anthrax and I thought 'oh my god the mayor is going to kill me' because he is already running late, and he has all these other things to do and he really doesn't want to hear about the life cycle of anthrax. But Giuliani completely engaged and they talked for hours… in discussion ranging from how many anthrax spores could fit in a light bulb to patterns of dispersal in the subway system potentially.[22]

Now that she had Giuliani's attention, they soon conducted "a tabletop exercise" — they played out an attack scenario with relevant parties to help understand the scope of the problem and devise strategies. As a result, New York City, unlike many other cities at the time, had a plan in place. So in March 1995 when the sarin nerve gas was released in the subway in Tokyo, Hamburg recalled that Giuliani "was perfectly positioned to say, 'I have already been briefed by the world's expert on this, and we've thought about it,' and he really embraced this… I'll never forget Josh and the life cycle of anthrax."[23] In terms of readiness for a bioweapons attack, New York was anomalous.

Hamburg then tried to get the Federal agencies prepared at the national level. She and Lederberg worked together again when President Clinton appointed her as Assistant Secretary for Planning and Evaluation in the Health Department. She wanted to establish a system that could detect a bioterrorism attack and mobilize a response at the national level, just as she had done in New York City. However, she received great resistance to the idea that such a system was needed. Some of her colleagues treated her fears as absurd fantasies.

Infectious disease threats were certainly part of her portfolio, but no one was thinking of bioterrorism when she arrived in Washington in 1997. Early recognition and response would have to come through public health programs in her view; it would take time to know that an attack had happened and prepare a response. But national defense of this kind was historically not a mandate of public health departments, and public health was historically not under the umbrella of law enforcement agencies. Anyway, to some, the whole idea of bioterrorism still seemed like alarmist fiction. Hamburg recalled the head of the National Institutes of Health saying to her, "You should be working on real world problems. You have been reading too many Tom Clancy novels."[24] The CIA also dismissed bioweapons and said terrorists would use explosives and knives.[25] The FBI believed that they could treat a biological weapons attack as they would a bomb threat — locate it and defuse it. Hamburg recalled, "I literally had an FBI agent say to me, we don't need you guys from public health because we would just go in and diffuse the pathogen."[26]

Neither the Department of Justice nor the Department of Defense was much interested in readiness for a disease outbreak, whether it was deliberately incited or not. As Hamburg put it, government officials would not think twice about spending a billion dollars to upgrade a nuclear submarine, which they hoped would never be used, but they were reluctant to allocate a billion dollars for systems to prevent a disease like the Zika virus, for example. They considered that a waste of money.[27] As Hamburg put it, it was important to have voices who can communicate the facts and identify how money should be invested: "So having people like Josh speaking to politicians, laying out scenarios, informing them of the realities about what has happened in the world was critical."[28]

Lederberg also knew how difficult it was getting government agencies to act to protect civilian populations against a biological weapons attack. He knew that biological weapons were widely available and there was no possibility of controlling its proliferation in states and substate groups because the cost and technology to make a bioweapon was low. A number of nations were alleged to have biowarfare programs of varying scale by the mid-1990s, including China, Iran, Iraq, Israel, Libya, North Korea, Syria, and Taiwan — as well as Russia.[29] The only recourse was deterrence and limiting damage after an attack. But the problem of civic preparedness for any bioweapons attack fell in the cracks between Federal agencies, and there was no coherent plan.

The Department of Defence was well informed of the bioweapons threat. In the spring of 1994, Lederberg was chairing a "Blue Ribbon Panel" surveilling the Army's research program in biological weapons defence which, in his view, was on a sound conceptual basis. However, civil defence was not part of the DOD's mandate. During Desert Shield, Lederberg had good conversations with George H. Bush's National Security Officer, who took the issue of biowarfare preparedness seriously. A "skeletal framework of coordination" with the relevant agencies had been put in place. But that all but disappeared after the Gulf War ended.

Lederberg explained all of this to Clinton's security advisor, Anthony Lake. They had met at a dinner in New York at the Council on Foreign Relations on June 11, 1994, a few weeks after he met with Giuliani. When Lederberg mentioned to him that the government was not doing enough to prepare for a biological weapons attack, Lake replied, "Do I really have to worry about this?"[30] Lederberg wrote him a letter the same day suggesting that he get "4 or 5 relevant people in one room to hear first hand what resources they could bring to bear" the next time there was an event comparable to the truck bomb attack on the World Trade Center, and there was even a question that a biological weapon was involved. "I lament, yes this is one more thing to worry about. But just because the responsibilities and technical capabilities cut across different agencies, I urge that it does deserve your attention."[31]

On April 10, 1998, Richard Clark, whom President Clinton had just appointed as National Coordinator for Security, arranged for Lederberg and a few other distinguished scientists and public health experts, including Margaret Hamburg, to brief the President on the need for a detection and response program to a bioweapons attack. At that time, the public health system in the US was in shambles compared to any other economically developed nation. Half of the health departments did not use e-mail and many had no access to Internet service. The Centers for Disease Control and Prevention's epidemiological surveillance system, created 50 years earlier to detect germ warfare attacks, was in ruins.[32]

Lederberg explained the scientific breakthroughs in the history of war, from the introduction of iron weapons to gun powder to the manipulation of genes for germ warfare that now gave rogue nations and terrorists an opportunity to strike deep into the nation.[33]

Clinton understood the urgency immediately. He knew that Iraq also had a biological weapons program that Saddam Hussein was unwilling to end. Hamburg recalled that Clinton sat them down after the meeting and recommended that they put in a request for a budget to really build a program in "security infrastructure and counterterrorism."[34]

At the end of the meeting, Lederberg gave Clinton a copy of the August 1997 issue of the *Journal of the American Medical Association*, which was devoted almost entirely to the threat of germ warfare. Lederberg had edited the issue and written its lead report, "Infectious Disease and Biological Weapons."[35] A White House staffer told him that Clinton read the scholarly articles and circulated the journal to his staff a few days later with his notes written in the margins. Lederberg smiled when he heard that, and as journalist Judith Miller and colleagues mused, "it was surely the first time a president had ever read an entire issue of the medical journal."[36]

Reacting to a biological bomb would be a complex process. The simplistic view that a vaccine was a cure-all to an epidemic was the conceit of the 1995 medical disaster film *Outbreak*, based on Richard Preston's *The Hot Zone*. The film actually begins with a quote from Lederberg: "The single biggest threat to man's continued dominance on this planet is the virus." But he thought the movie was simplistic and misleading. Much more than vaccines were needed for an epidemic. There was "no magic bullet," as Lederberg told the Senate Appropriations Committee on "civic preparedness" for bioterrorism in 1999:

> The public health infrastructure is the most important component; but this has to be designed and exercised to coordinate with all other elements of emergency management: public information, law enforcement and, if need be on such a scale, support from forces that can be mobilized under military discipline. Besides the structural arrangements would be provision for material: diagnostics, antidotes, hospital support equipment — including improvised beds, shelter and isolation, and so forth.[37]

Vaccines are only part of the required response to outbreaks, but they *are* an important part, and that meant further research on viruses. Lederberg had put up a cautionary flag when the World Health Organization declared in the 1970s that smallpox had been eradicated following its global immunization campaign.[38] Any talk of "eradication" could not be taken at face value, he argued.[39] Smallpox could live in the frozen arctic, for example, so the virus had to be kept available for research. Vaccination against it was critical because the build up of a large population of unvaccinated adults would be catastrophic in the event of a new epidemic.

Lederberg and Clinton had become close. When Clinton made the decision in 1999 to halt the proposed eradication of smallpox, Lederberg wrote an op-ed in *The Washington Post* to explain the reasoning behind it.[40] In January that year, Clinton announced his decision to ask Congress for 2.8 billion dollars to prepare for attacks with biological weapons and to combat computer-warfare threats as well.

With Lederberg at his side, Clinton gave a speech at the National Academy of Sciences to great applause in which he outlined his plan to increase spending on the nation's public health surveillance system and build a national stockpile of antibiotics and medicines against anthrax, smallpox, and pneumonic plague. As Miller and her journalist colleagues wrote: "The easy banter on the stage that day between the president and 'Josh' sent a powerful signal to the federal bureaucracy: Lederberg and his fellow advocates of biodefense had influence at the White House that could no longer be ignored."[41]

By that time, Lederberg had begun working for the Navy again, back where he had started his career as a young medical student during the Second World War. This time, though, he was a consultant working in the Defence Department with Richard Danzig, who was then Undersecretary of the Navy. The first issue they worked on in 1996 was whether the military should be vaccinated against anthrax, because there had been concerns since the Gulf War in 1990 that Iraq had bioweapons.[42] All three Joint Chiefs of Staff had initially voted against vaccination, but for different reasons. Danzig asked Lederberg to attend the meetings to address their concerns. Subsequently, they voted unanimously in support of it.[43]

Lederberg and Danzig worked together again when letters containing anthrax spores were sent to several news media offices and to two Democratic Senators in October 2001, following the destruction of the Trade Towers on September 11. Danzig was then Secretary of the Navy, but Scooter Libby, Chief of Staff for Vice-President Dick Cheney, asked him for help on bioterrorism and put together a plan of civilian defence. He and Lederberg were "synergistic," he said. Lederberg was different from the other Nobel Prize winners that Danzig knew, who "are seduced by

Fig. 29.1. Lederberg and Richard Danzig, Aug 2001.

it into thinking that their views on other subjects are particularly sanctified." Outside of biology, Lederberg "was just a participant in the discussion, a full-fledged one, but not one who regarded himself as empowered… Josh's qualities of character to me were every bit as important and relevant to his success as his qualities of mind… I think everyone thought that Josh was special."[44]

The Microbiome

On October 20, 2001, less than six weeks after the World Trade Towers fell and just after the anthrax letters, Lederberg gave a different kind of talk at the Centennial Celebration of the Nobel Prize in Medicine and Physiology hosted by the Robert Koch Institute in Berlin. It would not be on emerging diseases or bioterrorism this time; instead, he talked about another underdeveloped field of research — "the microbial communities living in and on us in various modes of conflict and cooperation." All the microbes that share our body space for better or for worse, "the microbiome", he said, should be thought of as part of our "extended genome" because microbes and host function as a "superorganism."[45]

In speaking about our microbiomes and ourselves as superorganisms, Lederberg was actually returning to arguments he had made 48 years earlier when he suggested that the concept of heredity had to be extended to include infectious genetic entities, and the boundaries of the organism had to be extended to embrace symbionts (Chapter 10).[46]

He returned to the superorganism as a concept for the new century in a paper on "Infectious History," which was published in *Science* in 2000. It was a chimeric overview — part history of science and part natural history — in which he explained why and how we need to learn to live with pathogens, and how simply aiming to kill them was quixotic. Microbes, he said, evolve at rates exceeding ours by a million fold, or perhaps even a billion fold. A year in microbial history would easily match the 200 million year span of all mammalian evolution. Our survival was a matter of "our wits versus their genes."

Adopting a "germ's eye view" of infections, Lederberg asked readers to turn away from the twentieth century's "metaphor of a war" in describing our relations with pathogens and to accept "a more ecologically informed metaphor."[47] One had to understand the ecological and social conditions that favor their proliferation: the human population explosion, crowding, poverty, destruction of forests, and routine long-distance travel of human hosts; indeed, Jules Verne's *Around the World in 80 Days* of 1873 could be rewritten as around the world in 80 hours.[48]

"Our most sophisticated leap," he said, "would be to drop the Manichaean view of microbes — 'We good; they evil.'"[49] Although microbes have a knack for making us ill and killing us, in the long run, microbes have a shared interest in their hosts' survival because "a dead host is a dead end for most pathogens." Evolution in the long run results in a co-adaptation of both. "We should think of each host and its parasites as a superorganism," he said, "with respective genomes yoked into a chimera of sorts."[50]

By this time, biology was much different than it had been five decades earlier when Lederberg last wrote about extending the boundaries of the organism to include infectious heredity. The fundamental importance of symbiosis in evolution had been fully documented with molecular evidence that mitochondria, the energy-generating organelles of cells, are vestiges of free-living bacteria that entered a cell billions of years ago. The same held true for chloroplasts, which harvest solar energy in plant cells and drive the production of oxygen and carbon that nourishes the rest of the biosphere.[51]

By the turn of the present century, molecular genetic studies of microbial evolution were also revealing a whole new vision of the microbiosphere as a worldwide web of informational exchange, with DNA as the packets of data going every which way throughout that world. This was no "superficial analogy", Lederberg said, while noting how many viruses can "download" their DNA into their host genomes. Many of our genes also originated from such viral encounters.[52]

There was also suggestive evidence that gut microbes play a role in immunity by warding off pathogens, as well as in digestion. There were even hints then that our gut microbes might have something to do with the increased prevalence of asthma.

But the fact remained, as Lederberg emphasized, that the microbes occupying our body surfaces, skin, gut, and mucous membranes were understudied; their functions were poorly understood. "Yet," he emphasized again, "they are equally part of the superorganism genome with which we engage the rest of the biosphere."[53]

In April 2001, Lederberg introduced the term "microbiome" in the journal founded by his old friend Eugene Garfield, *The Scientist,* to signify "the ecological community of commensal, symbiotic, and pathogenic microorganisms that literally share our body space which have been all but ignored as determinants of health and disease."[54] "Microbiome" is practically a household word today, and studies of our gut microbiomes have grown quickly during the first decades of this century.[55] Today, our microbiomes have been associated with various conditions, including asthma, diabetes, and autism. Had Lederberg lived to see it, I doubt he would have been surprised.

Chapter 30 An Extraordinary Life

> Knowing is not enough; we must apply.
> Willing is not enough; we must do.
>
> Goethe

Lederberg had the look of a distinguished polymath as he turned eighty in 2005. Slightly dishevelled, he spoke softly with his usual great precision and deliberation. "You sense not just intellectual power but moral force," a journalist commented. "There is something prophetic about him."[1] But he was not in good health. He had spinal stenosis, a narrowing of spaces in the spine that put pressure on the nerves. His pain required medication that rendered him tired and often unable to focus.

Earlier in October 2004, it had become impossible for him to work due to his back pain, and a few months later he opted for surgery. Unfortunately, his surgery was only partially successful and resulted in complications. He was in a wheelchair at his 80th birthday celebration at the Rockefeller, during which a roster of dignitaries and colleagues gave presentations addressing different aspects of his long and complex career. Another birthday party was organized for him at the National Academy of Sciences building in Washington.

Lederberg went to the White House on his last trip to Washington in December 2006. President George W. Bush honored

Fig. 30.1. George W. Bush awards the Presidential Medal of Freedom 2006. The Lederberg papers.

him with the nation's highest civil award, the Presidential Medal of Freedom. Bush stated:

> Joshua Lederberg has always seemed ahead of his time. He was researching genetics when the field was scarcely understood. He was studying the implications of space travel before there were astronauts. And even three decades ago, he was warning of the dangers of biological warfare. All of his life, people have seen something special in this rabbi's son from Montclair, New Jersey.

Josh died of pneumonia two years later on February 2, 2008.

As was typical of Lederberg, he had made arrangements for the remains of his life's work, but had given no thought to his funeral arrangements. He had worked with the National Library of Medicine to catalogue and digitize his correspondences and other archival material. By contrast, he gave his family no guidance whatsoever on what to do with his bodily remains. He had no patience for organized religion.

While agnostic, he had been deeply attached to his Judaic heritage. He had held Seders at his house during Passovers and invited close friends and colleagues. He presided over those events, read from Seder books in Hebrew, and explained what each passage meant.[2] When he donated a reprint of his Nobel Prize lecture to the Hebrew University Library in Jerusalem to add to others that he'd seen there in the summer of 1972, he included an inscription from Dante's Inferno (26): "Consider your origin; you were not born to live like brutes, but to follow virtue and knowledge."[3]

Lederberg was buried at Cedar Park across the Hudson River in Paramus, New Jersey in a cemetery plot associated with the *Or*

Fig. 30.2. Joshua Lederberg's gravesite. (Courtesy of Eugene Garfield.)

Fig. 30.3. Marguerite Lederberg and Eugene Garfield, at Gravesite, undated. (Courtesy of Eugene Garfield.)

Zarua synagogue that Marguerite belonged to.[4] Elie Wiesel was among those who delivered eulogies on a cold and rainy Tuesday morning in February. The flag at the Rockefeller was lowered.

The New York Times and *The Washington Post* summarized his scientific career and service to the nation. "He was one of the great scientists of the 20th century," science journalist William Broad wrote in *The Times*. "I know that's a strong statement, but it's justified."[5] Commentators covered his repertoire of achievements in bacterial genetics, immunology, and the foundations of molecular biology and biotechnology. He was a "brilliant analyst and visionary" who had pioneered artificial intelligence, explored extraterrestrial life, and advised governments across nine presidential administrations, over a half-century. "I know of no other eminent scientist who produced so much serious analysis of public policy and social problems, giving wise advice and stimulating new lines of inquiry. Our country and the world," his friend David Hamburg said, "are in his debt."[6]

Perhaps one of the most insightful private remarks was made by Lederberg's colleague and former president of the Rockefeller, Nobelist Torsten Wiesel. "You know, Josh was lucky," he told a colleague. "He got his Nobel Prize early so he could spend the rest of his life doing what he wanted."[7] Wiesel would not have known of Lederberg's misgivings about the Prize, of the talk of some colleagues around him that the award was premature, or of his going into a deep depression for several years after receiving it. Nonetheless, there can be no question that being endowed with the Nobel Prize at the young age of 33 permitted him to branch out in a way that matched his diverse intellectual interests.

Still, Wiesel's comment deserves more consideration because one cannot assume that receiving the Nobel Prize at an early age would work just as well for others. When Barbara McClintock, for example, was awarded the Nobel Prize in 1983 at the age of 81, she considered herself fortunate that the Prize came late in life because it entailed "a tremendous disruption of a very private life she had constructed for herself."[8] She had long been recognized as the most brilliant cytogeneticist in the world. Lederberg had nominated her for the Prize for her work on mobile genetic elements, the so-called "jumping genes" in corn, which she conducted decades earlier. It was not even her most important work, but the importance and generality of transposable elements was known by the 1980s; it was a hot topic, and one for which she could be awarded the Prize, which in her case was sort of a "lifetime achievement" award.[9]

While McClintock might have regretted an early loss of private scientific life, Lederberg seemed, in time at least, to relish it and he knew how to use it effectively. In 1967, when a journalist asked him straight up about how the Nobel Prize had changed his life, he replied: "Well, you are transformed from a private person to a public institution. Had I not received the prize, you surely would not know who I am nor would you wish to interview me." The prize, he said, "simply turns the spotlights of publicity on its recipient, and with that the matter more or less comes to an end."[10] Well, the matter certainly did not come to an end with that. This is not to say that the Nobel Prize led to his being regarded as a genius; all evidence suggests it was the other way around.

"Fortune favors the prepared mind," Louis Pasteur famously said. That is what it means to make an observation on a petri dish, and to be at the right place at the right time. Lederberg's luck began at Columbia University when, as a youthful prodigy, he won the support of his professors. Then he had the good fortune to work with Francis Ryan, fresh from his return from Stanford, and learn the methods of microbial genetics. And then he was offered the opportunity to work with Edward Tatum, who just happened to have the right bacterial mutants for the job at Yale.

As it turned out, even *E. coli* K-12 was lucky for him, because his demonstration of recombination would not have worked with many other strains. The meeting at Cold Spring Harbor in 1946, where the world's geneticists were introduced to the young man's genius, was also opportune. The timely offer of the position at the

University of Wisconsin where he was able to develop a research program was another stroke of very good luck. But it was Lederberg's genius, his collaborations with his first wife, Esther, his talented graduate students, and associates, and his elegantly simple experimental designs, such as replica plating and the use of antibiotics to find mutants for study, that brought him to prominence.

His Nobel Prize was fortuitously coupled with his move to Stanford's new medical faculty, with its rejection of the traditional trade school-style of medical training and embrace of fundamental biomedical research and interdisciplinarity. Stanford allowed him to move freely between specialties, exploit his diverse interests, and form new collaborations with talented chemists, computer scientists, psychiatrists, and social scientists. None of this was planned; all of it, in fact, diverged from his original plans.

After all, Lederberg had agreed to go to Stanford ostensibly to develop molecular biology and its medical application. His intention was to team up with Arthur Kornberg's biochemistry department. Despite his anxiety about taking on an administrative role as the founding chair of his own genetics department, and his "need" to get back in the laboratory, he really was a different kind of scientist. His skills were not really suited for the molecular biology laboratory. He orchestrated the research in his department, established new collaborations for his young colleagues, and pointed them to new fields to plow.

Lederberg's strength was in the breadth and depth of his knowledge and his ability to foresee and initiate new programs, think across disciplines, and embrace new ideas and take them to their logical consequences, even when they diverged from prevailing wisdom. That was his scientific signature which characterized his discovery of bacterial genetic recombination, development of the clonal selection theory of immunity, exobiology, and artificial intelligence, as well as prophetic warnings of emergent diseases and the imminent threat of biowarfare. The Nobel Prize gave him a voice.

Lederberg's eclecticism rubbed against the conception of the narrowly specialized researcher so characteristic of science then and now. Furthermore, that friction was not simply a matter of "generalist" versus "specialist." He was a polymath whose expert knowledge in diverse fields was apparent to everyone with whom he interacted in the university, government, industry, and the world at large.

It would be all too simple to suggest that Lederberg belonged to a generation of postwar American liberals who believed in the government's ability to improve society and ensure peace, and who believed in the responsibility of experts to help guide government action. Very few biologists took on that role, many looked askance at it, and undoubtedly few would have gained the trust that he garnered. His patriotic duty was heightened by his debt to the nation in fostering his education and keeping him off the battlefield during the war. In that regard, he never really abandoned the navy uniform he divested after the war — any more than he did medical school.

What is remarkable is how well-planned his way into policy truly was. It began in earnest after he came out of a period of deep depression and after his first marriage, when he started his weekly column on science and society in *The Washington Post*. His foray into advising the government as a scientific expert was not a matter of an empty niche waiting to be filled. It was a niche, like any other, that had to be created.

Untainted by partisan politics, Lederberg's advice was steeped in the scientific zeal of a humanist employing rationalism and critical thinking to the controversies that shape policy. His internationalism was embedded in his trust that the culture of science, with its underlying spirit of common purpose, could cut across the political and ideological conflicts that divide nations. He had a reverence for science, which offered not only a path toward personal enlightenment, but also formed a basis for rational action. He foresaw the inevitability of the present pandemic and the challenges of harnessing scientific research, advice, and public health policy measures to address it. Had he lived to see the global outbreak of Covid-19, "I told you so" would be the furthest thing from his mind.

Endnotes

1 The Polymath

1. Joshua Lederberg interview by Barbara Hyde, April 22, 1996, transcribed S/2002 courtesy of Pramod Srivistava 32pp., p. 1, The Joshua Lederberg Papers, Profiles in Science, US National Library of Medicine.
2. James F. Crow, "Joshua Lederberg, 1925–2008: A Tribute," *Genetics* 178 (2008): pp. 1139–1140; S. Gaylen Bradley, "Joshua Lederberg 1925–2008," *Biographical Memoirs of the National Academy of Sciences USA* 91 (2009): pp. 3–25; Walter Bodmer and Ann Ganesan, "Joshua Lederberg 23 May 1925–2 February 2008," *Biographical Memoirs of the Fellows of the Royal Society* 57 (2011): pp. 229–251.
3. See for example Laurie Garrett, *The Coming Plague. Newly Emerging Diseases in a World out of Balance* (New York: Farrar, Straus and Giroux, 1994); Judith Miller, Stephen Engelberg, and William Broad, *Germs. Biological Weapons and America's Secret War* (New York: Simon and Schuster, 2001).

2 Heretic

1. Joshua Lederberg, June 20, 1932, The Joshua Lederberg Papers, Profiles in Science, The US National Library of Medicine.
2. Joshua Lederberg interview by James Bohning, The Rockefeller University, New York, June 25, July 7, and December 9, 1992 (Philadelphia: Chemical Heritage Foundation, Oral History Transcript no. 0107), p. 3.
3. Raphael Bashan, "The Jewish Heritage of the Genetics Genius," Ma'ariv, November 3, 1967, unpublished, 10pp., p. 5, The Joshua Lederberg Papers.
4. Ibid., p. 2, p. 6
5. Joshua Lederberg interview by Bohning, p. 3.

6. Joshua Lederberg interview by Barry Teicher, 1998, Madison Wisconsin, 161pp., The Joshua Lederberg Papers.
7. Quoted in Walter Bodmer and Ann Ganesan, "Joshua Lederberg 23 May 1925–2 February 2008," *Biographical Memoirs of the Fellows of the Royal Society* 57 (2011): pp. 229–251, p. 232.
8. Lederberg interview by Bohning, pp. 12–13.
9. Ibid., p. 21.
10. Raphael Bashan, "The Jewish Heritage of the Genetics Genius," Ma'ariv, November 3, 1967, unpublished, 10pp., p. 5, The Joshua Lederberg Papers.
11. Ibid., p. 7.
12. Joshua Lederberg interview by Barbara Hyde, March 22, 1996, 32pp., p. 1., The Joshua Lederberg Papers.
13. Joshua Lederberg interview by Lev Pevner, March 20, 1996, The Nobel Prize Internet Archive.
14. Ibid.
15. Ibid.
16. Lederberg interview by Bohning, p. 11.
17. Lederberg interview by Pevner, p. 2.
18. Lederberg interview by Bohning, p. 14.
19. Ibid., p. 7.
20. Lederberg interview by Teicher, p. 25.
21. See Rena Subotnik, "Talented Developed: Conversations with Masters of the Arts and Sciences," *Journal for the Education of the Gifted* 18 (1995): pp. 210–226.
22. Joshua Lederberg, "Genetic Recombination in Bacteria: A Discovery Account," *Annual Review of Genetics* 21 (1987): pp. 23–46, p. 25.
23. Lederberg interview by Bohning, p. 13.
24. Ibid., p. 27.
25. Fanny S. Rippere to Joshua Lederberg, March 10, 1959, quoted in Joshua Lederberg and Harriet Zuckerman, "From Schizomycetes to Bacterial Sexuality: A Case Study of Discontinuity in Science," unpublished, The Joshua Lederberg Papers.
26. Lederberg interview by Bohning, p. 3.
27. Lederberg, "Genetic Recombination in Bacteria," p. 26.
28. Ibid.

29. Lederberg interview by Hyde, p. 2.
30. Lederberg, "Genetic Recombination in Bacteria," p. 25.
31. Joshua Lederberg interview by Bohning, pp. 25–26.
32. Ibid.

3 In Navy Uniform

1. Frank H. Stodola, "To Whom it May Concern," December 28, 1942, The Joshua Lederberg Papers, Profiles in Science, US National Library of Medicine.
2. Joshua Lederberg interview by James Bohning, The Rockefeller University, New York, June 25, July 7, and December 9, 1992 (Philadelphia, PA: Chemical Heritage Foundation, Oral History Transcript no. 0107), p. 13.
3. Ibid.
4. Joshua Lederberg, "Genetic Recombination in Bacteria: A Discovery Account," *Annual Review of Genetics* 21 (1987): pp. 23-46, p. 27.
5. Joshua Lederberg interview by Barry Teicher, Madison, Wisconsin, 1998, The Joshua Lederberg Papers.
6. W. Gordon Whaley to P. E. Haggerty, August 29, 1978, The Joshua Lederberg Papers.
7. Frank H. Stodola, "To Whom it May Concern," December 28, 1942, The Joshua Lederberg Papers.
8. Joshua Lederberg to Frank H. Stodola, Department of Organic Chemistry Columbia University, July 6, 1942, The Joshua Lederberg Papers.
9. Frank H. Stodola, "To Whom it May Concern," December 28, 1942, The Joshua Lederberg Papers.
10. Paul Marks interview by Jan Sapp, Palm Beach Florida, December 5, 2016.
11. Lederberg interview by Bohning, pp. 40–41.
12. Lederberg interview by Teicher.
13. Lederberg interview by Bohning, p. 42.
14. Ibid., pp. 39–40.
15. Joshua Lederberg, "Genetic Recombination in Bacteria: A Discovery Account," *Annual Review of Genetics* 21 (1987): pp. 23–46, p. 29.
16. Joshua Lederberg to Frank H. Stodola, Nov 4, 1944, The Joshua Lederberg Papers.

17. Lederberg to Jack Edman, August 3, 1945, The Joshua Lederberg Papers.
18. George Beadle, "Genes and Chemical Reaction in Neurospora," *Les Prix Nobel en 1958*, Stockholm 1959, pp. 147–159; Idem., "Biochemical Genetics: Some Recollections," in J. Cairns, G. S. Stent, and J. D. Watson, eds., *Phage and the Origins of Molecular Biology* (Cold Spring Harbor: Cold Spring Harbor Laboratory of Quantitative Biology, 1966), pp. 23–32, 66; Joshua Lederberg, "Edward Lawrie Tatum," *Biographical memoirs of the National Academy of Sciences* (1990), pp. 357–386.
19. George Beadle and Edward Tatum, "Genetic Control of Biochemical Reactions in *Neurospora*," *Proceedings of the National Academy of Sciences* 27 (1941): pp. 499–506.
20. Joshua Lederberg, "Francis Ryan," in Wesley First, ed., *University on the Heights*, (New York: Double Day, 1969), pp. 105–109, p. 105; Lederberg, "Genetic Recombination in Bacteria," p. 28.
21. Lederberg interview by Teicher, p. 14.
22. Ibid., p. 13.
23. Lederberg interview by Bohning, p. 44.
24. See for example H. F. Judson, *The Eighth Day of Creation. The Makers of the Revolution in Biology* (New York: Touchstone, 1979), p. 30; Robert Olby, *The Path to the Double Helix*; Lily Kay, *Who Wrote the Book of Life*, (Stanford University Press, 2000).
25. F. Griffith, "The Significance of Pneumococcal Types," *Journal of Hygiene* 27 (1928): pp. 113–159.
26. O. T. Avery, C. M. MacLeod, and M. McCarty, "Studies on the Chemical Nature of the Substance Inducing Transformation of Pneumococcal Types," *Journal of Experimental Medicine* 79 (1944): pp. 137–158.
27. See Joshua Lederberg, "The Transformation of Genetics by DNA: An Anniversary Celebration of Avery, Macleod And McCarty (1944)," *Genetics* 136 (1994): pp. 423–426.
28. Lederberg diary note, January 20, 1945, The Joshua Lederberg Papers.
29. A. E. Mirsky and A. W. Pollister, "Chromosin, a Desoyribose Nucleoprotein Complex of the Cell Nucleus," *Journal of General* Physiology 30 (1946): pp. 117–148; G. W. Beadle, "Genes and Biological Enigmas," *American Scientist* 36 (1948): pp. 71–74; Mathew Cobb, "Oswald Avery, DNA, and the Transformation of Biology," *Current Biology* 24 (2014): pp. R55–R60.

30. Lederberg, "Genetic Recombination in Bacteria: A Discovery Account."
31. Joshua Lederberg, "Francis Ryan," pp. 105–109, in Wesley First, ed., *University on the Heights* (New York: Double Day, 1969), p. 108.

4 Lucky

1. Joshua Lederberg to Edward Tatum, September 19, 1945, The Joshua Lederberg Papers, Profiles in Science, The US National Library of Medicine.
2. Jan Sapp, *The New Foundations of Evolution: On the Tree of Life* (New York: Oxford University Press 2009), p. 42.
3. Ibid.
4. Ibid.
5. Julian Huxley, *Evolution: The Modern Synthesis* (London: Allen and Unwin, 1942), pp. 131–132; Joshua Lederberg, "Problems in Microbial Genetics," *Heredity* 11 (1948): pp. 145–198, p. 153.
6. Joshua Lederberg, "Genetic Recombination in Bacteria: A Discovery Account," *Annual Review of Genetics* 21 (1987): pp. 23–46, p. 34.
7. Ibid.
8. F. J. Ryan and J. Lederberg, "Reverse Mutation and Adaptation in Leucine-less *Neurospora*," *Proceedings of the National Academy of Sciences USA* 32 (1946): pp. 163–173.
9. George Beadle to Joshua Lederberg, September 17, 1945, The Joshua Lederberg Papers.
10. Ibid.
11. Edward Tatum, "X-Ray Induced Mutant Strains of *Escherichia coli*," *Proceedings of the National Academy of Sciences USA* 31 (1945): pp. 215–219, p. 215; See also C. H. Gray and E. Tatum, "X-ray Induced Growth Factor Requirements in Bacteria," *Proceedings of the National Academy of Sciences USA* 30 (1944): pp. 404–410, p. 409; E. L. Tatum and David Bonner, "Indole and Serine in the Biosynthesis and Breakdown of Tryptophan," *Proceedings of the National Academy of Sciences USA* 30 (1944): pp. 30–37.
12. E. L. Tatum and Joshua Lederberg, "Gene Recombination in the Bacterium *Escherichia coli*," *Journal of Bacteriology* 53 (1947): pp. 673–684, pp. 674–675.

13. Joshua Lederberg to Edward Tatum, September 19, 1945, The Joshua Lederberg Papers.
14. Joshua Lederberg interview by James Bohning at the Rockefeller University, New York, June 25, July 7, and December 9, 1992 (Philadelphia: Chemical Heritage Foundation, Oral History Transcript no. 0107), p. 47.
15. Edward Tatum to Joshua Lederberg, October 15, 1945, The Joshua Lederberg Papers.
16. Lederberg to Edward Tatum, November 26, 1945, The Joshua Lederberg Papers.
17. Tatum to Lederberg, December 7, 1945, The Joshua Lederberg Papers.
18. Joshua Lederberg, "A Nutritional Concept of Cancer," *Science* 104 (1946): p. 428.
19. Joshua Lederberg to Aura E. Sevringhaus, December 30, 1945, The Joshua Lederberg Papers.
20. Lederberg interview by James Bohning, pp. 48–49.
21. Joshua Lederberg, "Genetic Recombination in Bacteria: A Discovery Account," *Annual Review of Genetics* 21 (1987): pp. 23–46, p. 35.
22. Lederberg interview by Bohning, p. 30.
23. Joshua Lederberg interview by Barry Teicher, June 19, 1998, Madison Wisconsin, The Joshua Lederberg Papers, p. 20.
24. Joshua Lederberg, "Genetic Recombination in Bacteria: A Discovery Account," p. 35.
25. Lederberg interview by Teicher, p. 20–21.
26. Joshua Lederberg, "Genetic Recombination in *Escherichia coli*: Disputation at Cold Spring Harbor, 1946–1996," *Genetics* 144 (1996): pp. 439–443.
27. Seymour Hutner to Joshua Lederberg, April 16, 1992, The Joshua Lederberg Papers.
28. David Hamburg interview by Jan Sapp, The Nuclear Threat Initiative, Washington DC, May 20, 2016.
29. André Lwoff to Joshua Lederberg, January 3, 1985, The Joshua Lederberg Papers.
30. Joshua Lederberg, "Genetic Recombination in *Escherichia coli*: Disputation at Cold Spring Harbor," p. 441.
31. Ibid.
32. André Lwoff to Joshua Lederberg, January 3, 1985, The Joshua Lederberg Papers.

33. Joshua Lederberg and Edward Tatum, "Novel Genotypes in Mixed Cultures of Biochemical Mutants of Bacteria," *Cold Spring Harbor Symposium for Quantitative Biology* (1946): pp. 113–114, p. 114.
34. Joshua Lederberg interview by Sharon Zane, University of Wisconsin, March 26, 1998, pp. 22–23, The Joshua Lederberg Papers.
35. M. Zelle and J. Lederberg, "Single Cell Isolation of Diploid Heterozygous *Escherichia coli*," *Journal of Bacteriology* 61 (1951): pp. 351–355.

5 The Road Not Taken

1. Francis Ryan to Joshua Lederberg, August 2, 1946, The Joshua Lederberg Papers, Profiles in Science, The US National Library of Medicine.
2. Joshua Lederberg and Edward Tatum, "Gene Recombination in *Escherichia coli*," *Nature* 158 (1946): p. 558.
3. George Beadle to Edward Tatum, September 9, 1946, The Joshua Lederberg Papers.
4. Sol Spiegelman to Joshua Lederberg, November 11, 1946, The Joshua Lederberg Papers.
5. Joshua Lederberg to Sol Spiegelman, November 14, 1946, The Joshua Lederberg Papers.
6. Abraham Penzer to Joshua Lederberg, October 7, 1946, The Joshua Lederberg Papers.
7. Joshua Lederberg interview by James Bohning, The Rockefeller University, New York, June 25, July 7, and December 9, 1992 (Philadelphia: Chemical Heritage Foundation, Oral History Transcript no. 0107).
8. Norman Giles and Esther Zimmer Lederberg, "Induced Reversions of Biochemical Mutants in *Neurospora Crassa*," *American Journal of Botany* 35 (1948): pp. 150–157; Mary E. Case and Frederick J. de Serres, "Norman Henry Giles 1915–2006," *Biographical Memoirs of the National Academy of Sciences* (2007): pp. 3–17.
9. Joshua Lederberg interview by Bohning, p. 67.
10. See Joshua Lederberg, "Autobiographical Notes re: First Wife Esther Zimmer," December 31, 1985, pp. 1–13, The Joshua Lederberg Papers.
11. Joshua Lederberg to A. E. Sevringhaus, June 21, 1946; Joshua Lederberg to A. E. Sevringhaus, September 4, 1946; A. E. Sevringhaus to Joshua Lederberg, September 10, 1946, The Joshua Lederberg Papers.

12. A. D. Hershey, "Spontaneous Mutations in Bacteria Viruses," *Cold Spring Harbor Symposium for Quantitative Biology* 11 (1946): pp. 67–77.
13. S. E. Luria and M. Delbrück, "Mutations of Bacteria from Virus Sensitivity to Virus Resistance," *Genetics* 28 (1943): pp. 491–511, p. 492.
14. Max Delbrück to J. B. S. Haldane, August 20, 1946, The Joshua Lederberg Papers.
15. Quoted in William Lanuouette, *Genius in the Shadows: A Biography of Leo Szilard* (New York: Skyhorse Publishing, 2013), Chapter 25.
16. Joshua Lederberg to Leo Szilard, June 3, 1948, The Joshua Lederberg Papers.
17. E. L. Tatum and J. Lederberg, "Gene Recombination in the Bacterium *Escherichia coli*," *Journal of Bacteriology* 53 (1947): pp. 673–684.
18. S. E. Luria to Joshua Lederberg, March 26, 1947, The Joshua Lederberg Papers.
19. His PhD from Yale was granted June 1948, Official transcript of Record, Yale University Graduate School 22 February, 1949, Joshua Lederberg Papers, National Institutes of Health, Archives Box 223, Folder 35.
20. E. L. Tatum and J. Lederberg, "Gene Recombination in the Bacterium *Escherichia coli*," *Journal of Bacteriology* 53 (1947): pp. 673–684.
21. Salvador Luria to Joshua Lederberg, May 5, 1947, The Joshua Lederberg Papers.
22. Ralph Meader to Aura E. Severinghaus, May 15, 1947, The Joshua Lederberg Papers.
23. Aura E. Severinghaus to Ralph Meader, May 16, 1947; Ralph Meader to Aura E. Severinghaus, May 21, 1947, The Joshua Lederberg Papers.

6 Personality Matters

1. Edward Tatum to R. A. Brink, May 12, 1947, Joshua Lederberg file, Box 9, The University of Wisconsin Archives, Madison Wisconsin.
2. R. A. Brink to Edward Tatum, April 30, 1947, Joshua Lederberg file, Box 9, The University of Wisconsin Archives.
3. Edward Tatum to R. A. Brink, May 12, 1947, Joshua Lederberg file, Box 9, The University of Wisconsin Archives.
4. Edward Tatum to R. A. Brink, May 24, 1947, Joshua Lederberg file, Box 9, The University of Wisconsin Archives.

5. R. Alexander Brink interview by Donna S. Taylor, University of Wisconsin Madison, 1982, University Archives Oral History Project, pp. 40–42.
6. Ibid., p. 46.
7. Joshua Lederberg, "Gene Recombination and Linked Segregation in *Escherchia coli*," *Genetics* 32 (1947): pp. 505–525.
8. Sol Spiegelman to Joshua Lederberg, October 15, 1947, The Joshua Lederberg Papers, Profiles in Science, The US National Library of Medicine.
9. R. A. Brink to Edward Tatum, June 25, 1947, Joshua Lederberg file, Box 9, The University of Wisconsin Archives.
10. Polly Bunting to Joshua Lederberg, August 29, 1977, The Joshua Lederberg Papers.
11. Stanhope Bayne-Jones to Ira Baldwin, July 14, 1947, Joshua Lederberg file, Box 9, The University of Wisconsin Archives.
12. Dennis Watson to R. A. Brink, July 10, 1947, The Joshua Lederberg Papers.
13. Dennis Watson to R. A. Brink, July 10, 1947, The Joshua Lederberg Papers.
14. Edward Tatum to R. A. Brink, May 12, 1947, Joshua Lederberg file, Box 9, The University of Wisconsin Archives.
15. Edward Tatum to R. A. Brink, May 12, 1947, Joshua Lederberg file, Box 9, The University of Wisconsin Archives.
16. Ibid.
17. Edmund W. Sinnott to R. A. Brink, undated letter, June 1947, Joshua Lederberg file, Box 9, The University of Wisconsin Archives.
18. Ibid.
19. Ibid., p. 48.
20. R. A. Brink to Joshua Lederberg, July 19, 1947, The Joshua Lederberg Papers.
21. Joshua Lederberg interview by James Bohning, The Rockefeller University, New York, June 25, July 7, and December 9, 1992 (Philadelphia: Chemical Heritage Foundation, Oral History Transcript no. 0107).
22. R. A. Brink to Max Zelle, August 25, 1947, The Joshua Lederberg Papers.
23. Ibid.
24. Joshua Lederberg interview by Barry Teicher, Madison Wisconsin, 1998, p. 40. The Joshua Lederberg Papers.
25. Ibid., p. 53
26. Brink Interview by Donna S. Taylor, University of Wisconsin Oral History Project, Madison 1982, pp. 50–51, The Joshua Lederberg Papers.

27. Brink Interview by Donna S. Taylor, p. 50.
28. James F. Crow interview by Laura L. Smail, The University of Wisconsin-Madison Archives Oral History Project, 1983, p. 36, The Joshua Lederberg Papers.
29. Brink Interview by Taylor, pp. 49–50.
30. Joshua Lederberg to R. A. Brink, September 10, 1973, The Joshua Lederberg Papers.
31. Ibid., pp. 40–41.
32. See for example Lily E. Kay, *The Molecular Vision of Life: Caltech, the Rockefeller Foundation, and the Rise of the New Biology*, (New York: Oxford University Press, 1996).
33. R. A. Brink to Warren Weaver, December 4, 1947, 200D Genetics, Lederberg, Joshua, 1947–1952, The Rockefeller Foundation Archives.
34. Weaver Diary note, January 19, 1948, 200D Genetics, Lederberg, Joshua, 1947–1952, The Rockefeller Foundation Archives.
35. Warren Weaver diary of interview, May 12, 1949, 200D Genetics, Lederberg, Joshua, 1947–1952, The Rockefeller Foundation Archives; W. F. Loomis Diary, January 16, 1950, 200D Genetics, Lederberg, Joshua, 1947–1952, The Rockefeller Foundation Archives.

7 A Field of Their Own

1. Joshua Lederberg interview by Barry Teicher, Madison Wisconsin, 1998, p. 59, The Joshua Lederberg Papers, Profiles in Science, The US National Library of Medicine.
2. Norton Zinder memoir, June 7, 2004, The Joshua Lederberg Papers.
3. Norton Zinder, undated untitled manuscript, p. 14, The Joshua Lederberg Papers.
4. See for example Katy Steinmetz, "Esther Lederberg and Her Husband were Both Trailblazing Scientists. Why have More People Heard of Him?" *Time*, April 11, 2019.
5. R. A. Brink interview by Donna S. Taylor, The University of Wisconsin Archives Oral History Project, 1982; R. A. Brink Papers, The University of Wisconsin, Madison, p. 51.

6. See for example Katy Steinmetz, "Esther Lederberg and Her Husband were Both Trailblazing Scientists. Why have More People Heard of Him?" *Time*, April 11, 2019.
7. See Joshua Lederberg, "Replica Plating and Indirect Selection of Bacterial Mutants: Isolation of Preadaptive Mutants in Bacteria by Sib Selection," *Genetics* 121 (1989): pp. 395–399.
8. J. Lederberg and E. M. Lederberg, "Replica Plating and Indirect Selection of Bacterial Mutants," *Journal of Bacteriology* 63 (1952): pp. 399–406, p. 400.
9. Quoted in S. Gaylen Bradley, "Joshua Lederberg 1925–2008," *Biographical Memoirs of the National Academy of Sciences* 91 (2009): pp. 1–25, p. 15.
10. Harry Eagle to Joshua Lederberg, October 10, 1951, The Joshua Lederberg Papers.
11. See for example Katy Steinmetz, "Esther Lederberg and Her Husband were Both Trailblazing Scientists. Why have More People Heard of Him?" *Time*, April 11, 2019.
12. Lederberg interview by Barry Teicher, p. 88.
13. Esther Lederberg, "Lysogenicity in *Escherichia coli* strain K-12," *Microbial Genetics Bulletin* 1 (1950), pp. 5–8; Esther M. Lederberg, "Lysogenicity in *B. coli* K-12," *Genetics* 86 (1951): p. 560.
14. Esther Lederberg and Joshua Lederberg, "Genetic Studies of Lysogencity in *Escherichia coli*," *Genetics* 38 (1953): pp. 51–64.
15. William Summers, *Félix d'Hérelle and the Origins of Molecular Biology* (New Haven: Yale University Press, 1999); Jan Sapp, *Evolution by Association. A History of Symbiosis* (New York: Oxford University Press, 1994).
16. See Max Gottesman and Robert Weisberg, "Little Lambda, Who Made Thee?" *Microbiology and Molecular Biology Reviews* 68 (2004): pp. 796–813.
17. Joshua Lederberg, "Genetic Recombination in Bacteria: A Discovery Account," *Annual Review of Genetics* 21 (1987): pp. 23–46, p. 33; Joshua Lederberg interview by Barry Teicher, University of Wisconsin, 1998, p. 27, The Joshua Lederberg Papers, The University of Wisconsin-Madison Archives.
18. Norton Zinder to Joshua Lederberg, June 16, 1948, The Joshua Lederberg Papers, Box 8, Folder 142, National Institutes of Health Archives, National Library of Medicine, Bethesda, Maryland.

19. Francis Ryan to Joshua Lederberg, June 22, 1948, The Joshua Lederberg Papers, The US National Library of Medicine, Profiles in Science.
20. Lederberg interview by Teicher, p. 71.
21. G. L. Hobby, K. Meyer, and E. Chaffee, "Activity of Penicillin in Vitro," *Proceedings of the Society for Experimental Biology and Medicine* 50 (1942): pp. 281–285.
22. Zinder memoire, 2004, unpublished, p. 2, The Joshua Lederberg Papers.
23. J. Lederberg and N. Zinder, "Concentration of Biochemical Mutants of Bacteria with Penicillin," *Journal of the American Chemical Society* 70 (1948): p. 4267.
24. Harry Eagle to Joshua Lederberg, November 7, 1956, The Joshua Lederberg Papers.
25. Karre Lilleengen to Joshua Lederberg, February 2, 1950, The Joshua Lederberg Papers.
26. Norton Zinder, "Forty Years Ago: The Discovery of Bacterial Transduction," *Genetics* 132 (1992): pp. 291–294.
27. Bernard Davis, "Nonfilterability of the Agents of Recombination in *Escherichia coli*," *Journal of Bacteriology* 60 (1950): pp. 507–508.
28. Emma Klieneberger-Nobel, "Filterable Forms of Bacteria," *Bacteriological Reviews* 15 (1951): pp. 77–103; J. Lederberg, Esther M. Lederberg, N. D., Zinder, and Ethelyn R. Lively, "Recombination analysis of bacterial heredity," *Cold Spring Harbor Symposia on Quantitative Biol*ogy 16 (1951): pp. 413–443, p. 438.
29. See Evelyn Fox Keller, *A Feeling for the Organism* (New York: Macmillan, 1984); Nathanal Comfort, *The Tangled Field* (Cambridge: Harvard University Press, 2001).
30. Zinder, "Forty Years Ago," p. 293.
31. Zinder, "Forty years Ago," p. 293.
32. Charlotte Auerbach to Joshua Lederberg, May 29, 1979, The Joshua Lederberg Papers.
33. Arnold W. Ravin, "Harriett Ephrussi-Taylor," *Genetics* 60 (1968): p. 524.
34. Zinder, "Forty years Ago", p. 293.
35. Ibid., p. 293.
36. Norton Zinder and Joshua Lederberg, "Genetic Exchange in Salmonella," *Journal of Bacteriology* 64 (1952): pp. 679–699, p. 679.

37. Evelyn Witkin, "Remembering Rollin Hotchkiss, 1911–2004" *Genetics* 170 (2005): pp. 1443–1447.
38. R. D. Hotchkiss, "Transfer of Penicillin Resistance in Pneumococci by the Desoxyribonucleate Derived from Resistant Cultures," *Cold Spring Harbor Symposium on Quantitative Biology* 16 (1951): pp. 457–461.
39. Joshua Lederberg and Philip Edwards, "Serotypic Recombination in Salmonella," *Journal of Immunology* 71 (1953): pp. 232–240; Joshua Lederberg and Tetsuo Iino, "Phase Variation in Salmonella," *Genetics* 41 (1956): pp. 743–757.
40. M. L. Morse, E. M. Lederberg, and J. Lederberg, "Transduction in *Escherichia coli* K-12," *Genetics* 41 (1956): pp. 142–156; Idem., "Transductional Heterogenotes in *Escherichia coli*," *Genetics* 41 (1956): pp. 758–779.

8 Sex Controversy

1. James Watson, *The Double Helix* (London: Readers Union Weidenfeld and Nicolson, 1968), pp. 77–78.
2. Joshua Lederberg, "Gene Recombination and Linked Segregations in *Escherichia coli*," *Genetics* 32 (1947): pp. 505–525, p. 505.
3. Max Delbrück to Joshua Lederberg, February 26, 1948, The Joshua Lederberg Papers, Profiles in Science, The US National Library of Medicine.
4. Joshua Lederberg, "Aberrant Heterozygotes in *Escherichia coli*," *Proceedings of the National Academy of Sciences* 35 (1949): pp. 178–184; Idem., "Bacterial Variation," *Annual Review in Microbiology* 3 (1949): pp. 1–22.
5. Lederberg, "Gene Recombination and Linked Segregations in *Escherichia coli*," p. 523.
6. Ibid.
7. L. L. Cavalli-Sforza, "Forty Years Ago in Genetics: The Unorthodox Mating Behavior of Bacteria," *Genetics* 132 (1992): pp. 635–637, p. 635.
8. Ibid., p. 635.
9. Joshua Lederberg to Luca Cavalli, October 7, 1948, The Joshua Lederberg Papers.
10. Cavalli-Sforza, "Forty Years Ago in Genetics," p. 635.
11. Ibid., p. 636.
12. Simon Silver *et al.*, "William Hayes: Pioneering Contributions Remembered," *ASM News* 61 (1995): pp. 17–20, p. 17; R. Jayaraman, "William

Hayes and His Pallanza Bob Shell," *Resonance* (2011): pp. 911–921, p. 911; Thomas D. Brock, *The Emergence of Bacterial Genetics* (Cold Spring Harbor: Cold Spring Harbor Laboratory Press, 1990), p. 91.
13. William Hayes, Autobiography, unpublished, 1986, p. 17, The Joshua Lederberg Papers; Thomas Brock, *The Emergence of Bacterial Genetics*, p. 89.
14. W. Hayes, "Recombination in *Bact. coli K* 12: Unidirectional Transfer of Genetic Material," *Nature* 169 (1952): pp. 118–119, p. 119.
15. Ibid., 119. See also W. Hayes, "The Mechanism of Genetic Recombination in *Escherichia coli*," *Cold Spring Harbor Symposia on Quantitative Biology* 18 (1953): pp. 75–93; See also William Hayes, "Genetic Recombination in *Bact coli K*12: Analysis of the Stimulating Effect of Utra-violet Light," *Nature* (1952): pp. 1017–1018.
16. William Hayes to Joshua Lederberg, December 3, 1951, The Joshua Lederberg Papers.
17. E. M. Lederberg, "Lysogencity in E. coli K-12," *Genetics* 36 (1951): p. 560.
18. Joshua Lederberg to William Hayes, January 7, 1952, The Joshua Lederberg Papers.
19. Luca Cavalli to William Hayes, February 16, 1952, The Joshua Lederberg Papers.
20. William Hayes to Luca Cavalli, February 20, 1952, The Joshua Lederberg Papers.
21. J. Lederberg, L. L. Cavalli, and E. M. Lederberg, "Sex Compatibility in *Escherichia coli*," *Genetics* 37 (1952): pp. 720–730.
22. See J. Lederberg, L. L. Cavalli, and E. M. Lederberg, "Sex Compatibility in *Escherichia coli*," *Genetics* 37 (1952): pp. 720–730, p. 725; See also William Hayes to Luca Cavalli, April 2, 1952, The Joshua Lederberg Papers.
23. Luca Cavalli to William Hayes, April 4, 1952, The Joshua Lederberg Papers.
24. William Hayes, "Observations on a Transmissible Agent Determining Sexual Differentiation in Bacterium *coli*." *Journal of General Microbi*ology 8 (1953): pp. 72–88, p. 72.
25. L. L. Cavalli, J. Lederberg, and E. M. Lederberg, "An Infective Factor Controlling Sex Compatibility in *Bacterium coli*," *Journal of General Microbiology* 8 (1953): pp. 89–103.

26. James D. Watson, *The Double Helix*, p. 77.
27. Brenda Maddox, *Rosalind Franklin: The Dark Lady of DNA* (New York: Harper Collins, 2002).
28. J. D. Watson to Francis Crick, November 3, 1966, The Joshua Lederberg Papers.
29. Joshua Lederberg to Robert K. Merton, November 9, 1971. The Joshua Lederberg Papers; See also Lederberg, "Genetic Recombination in Bacteria: A Discovery Account," *Annual Review of Genetics* 21 (1987): pp. 23–46, p. 42.
30. Hayes, Autobiography, p. 18.
31. William Hayes to Esther and Joshua Lederberg, September 21, 1953. The Joshua Lederberg Papers.
32. Hayes, Autobiography, p. 23.
33. Watson, *The Double Helix,* p. 77.
34. Ibid., p. 78.
35. J. D. Watson and W. Hayes, "Genetic Exchange in *Escherichia coli* K12: Evidence for Three Linkage Groups," *Proceedings of the National Academy of Sciences USA* 39 (1953): pp. 416–426.
36. Joshua Lederberg to Hayes, April 14, 1952, The Joshua Lederberg Papers, Profiles in Science.
37. Luca Cavalli to Joshua Lederberg, August 19, 1949, The Joshua Lederberg Papers; J. Lederberg, L. L. Cavalli, and E. M. Lederberg, "Sex Compatibility in *Escherichia coli,*" *Genetics* 37 (1952): pp. 720–730, p. 725.
38. F. Jacob and E. L Wolman, "Étape de la recombinaison génétique chez Escherichia coli K 12," *Comptes rendus Académie des Sciences* 240 (1956): pp. 2566–2568; E. Wollman and F. Jacob, "Sur le processus de conjugaison et de recombinaison chez," *Escherichia coli. Annales d'Institute Pasteur* 91 (1956): pp. 486–510.
39. F. Jacob and E. L. Wollman, "Recombination génétique et mutants de fertilité chez *Escherchia coli,*" *Compte rendus Académie des Sciences* 242 (1956): pp. 303–306.
40. Hayes, Autobiography, p. 20.
41. Joshua Lederberg interview by Barry Teicher, University of Wisconsin Archives Oral History Project, June 1998, p. 64.
42. Ibid., pp. 63–64.

9 The Extended Genotype

1. Joshua Lederberg, "Cell Genetics and Hereditary Symbiosis," *Physiological Reviews* 32 (1952): pp. 403–430, p. 403.
2. Ibid.
3. See Jan Sapp, *Evolution by Association. A History of Symbiosis* (New York: Oxford University Press, 1994).
4. Ibid.
5. E. B. Wilson, *The Cell in Development and Heredity* (New York: Macmillan, 1925), p. 39.
6. Sapp, *Evolution by Association.*
7. Joshua Lederberg, "Problems in Microbial Genetics," *Heredity* 2 (1948): pp. 145–198, p. 182; See also Joshua Lederberg, "Genetic Variation," *Annual Review of Microbiology* 2 (1949): pp. 1–22, p. 18.
8. E. M. Lederberg and J. Lederberg, "Genetic Studies of Lysogeny in *Escherichia coli*," *Genetics* 38 (1953): pp. 51–64, p. 61.
9. Gunther Stent, *Molecular Biology of Bacterial Viruses* (San Francisco: W. H. Freeman, 1963), p. 14; See also Horace Judson, *The Eighth Day of Creation: The Makers of the Revolution in Biology* (New York: Simon and Schuster, 1979), p. 373.
10. Joshua Lederberg, "Genetic Studies in Bacteria," in L. C. Dunn, ed., *Genetics in the 20th Century* (New York: Macmillan, 1951), pp. 263–290, p. 286.
11. Joshua Lederberg, "Cell Genetics and Hereditary Symbiosis," *Physiological Reviews* 32 (1952): pp. 403–430, p. 403.
12. Ibid., pp. 403–404.
13. Ibid., p. 424.
14. Ibid., p. 425.
15. Ibid.
16. Ibid., p. 415.
17. Ibid., p. 424.
18. Ibid., pp. 422–423.
19. Ibid., p. 426.
20. Ibid., pp. 426–427.
21. Michael Dietrich, "Richard Goldschmidt: Hopeful Monsters and Other 'Heresies,'" *Nature Reviews Genetics* 4 (2003): pp. 68–74.

22. Richard Goldschmidt to Joshua Lederberg, April 25, 1953, The Joshua Lederberg Papers, Profiles in Science, The US National Library of Medicine.
23. Sapp, *The New Foundations of Evolution*.
24. René Dubos to Joshua Lederberg, January 2, 1958, The Joshua Lederberg Papers.
25. Joshua Lederberg to René Dubos, January 7, 1958, The Joshua Lederberg Papers.
26. René Dubos and Russell Schaedler, "The Effects of the Intestinal Flora on the Growth and Rate of Mice, and on their Susceptibility to Experimental Infections," *Journal of Experimental Medicine* 111 (1960): pp. 407–417; Russell Schaedler and René Dubos, "The Susceptibility of Mice to Bacterial Endotoxins," *Journal of Experimental Medicine* 113 (1961): pp. 559–570; René Dubos and Russell Schaedler, "The Digestive Tract as an Ecosystem," *American Journal of Medical Sciences* 248 (1964): pp. 267–272.
27. R. Dubos, "Integrative and Creative Aspects of Infection," pp. 200–205, in M. Pollard, ed., *Perspectives in Virology*, Vol 2 (Minneapolis: Burgess Publishing, 1961).
28. Ibid., p. 204.
29. Mary Barber, "Coagulase-Positive Staphylococci Resistant to Penicillin," *Journal of Pathology and Bacteriology* 59 (1947): pp. 373–384.
30. Mary Barber, "Hospital Infection, Yesterday and Today," *Journal of Clinical Pathology* (London) 14 (1961): pp. 2–10, p. 4.
31. T. D. Brock, *The Emergence of Bacterial Genetics*. (Cold Spring Harbor: Cold Spring Harbor Laboratory Press, 1990), p. 107.
32. T. Watanabe, "Infective Heredity of Multiple Drug Resistance in Bacteria," *Bacteriological Reviews* 27 (1963): pp. 87–115, p. 87.
33. Lederberg, "Hereditary Symbiosis," p. 424.
34. Sapp, *The New Foundations of Evolution*.
35. See for example, Konstantin Khodosevich, L. Lebedev, and E. Sverdolv, "Endogenous Retroviruses and Human Evolution," *Comparative and Functional Genomics* 3 (2002): pp. 494–498; F. P. Ryan, "Human Endogenous Retroviruses in Health and Disease: A Symbiotic Perspective," *Journal of the Royal Society of Medicine* 97 (2004): pp. 560–565.
36. Scott Gilbert *et al.*, "A Symbiotic View of Life: We have Never been Individuals," *Quarterly Review of Biology* 87 (2012): pp. 325–341; Kevin Theis,

et al. "Getting the Hologenome Concept Right: An Eco-Evolutionary Framework for Hosts and Their Microbiomes," *mSystems* 1(2) (2012): e00028–16.

10 Down Under Immunity

1. Joshua Lederberg interview by Berry Teicher, Madison Wisconsin, 1998, p. 109, The Joshua Lederberg Papers, Profiles in Science, US National Library of Medicine.
2. Joshua Lederberg, "Genes and Antibodies," *Science* 129 (1959): pp. 1649–1653.
3. See Joshua Lederberg to Macfarlane Burnet, January 11, 1957; Sydney Rubbo to Joshua Lederberg, February 6, 1957, The Joshua Lederberg Papers.
4. F. M. Burnet, *Virus as Organism* (Cambridge, Mass: Harvard University Press, 1945); Idem., *The Principles of Animal Virology* (New York: Academic Press, 1950).
5. Lederberg interview by Teicher, p. 91.
6. Burnet to Joshua Lederberg, December 12, 1947, The Joshua Lederberg Papers.
7. Joshua Lederberg and T. Iino, "Phase Variation in Salmonella," *Genetics* 41 (1956): pp. 743–757; Lederberg interview by Teicher, p. 177.
8. Jacques Monod and François Jacob, "Genetic Regulatory Mechanisms in the Synthesis of Proteins," *Journal of Molecular Biology* 3 (1961): pp. 318–356.
9. Joshua Lederberg, "Ontogeny of the Clonal Selection Theory of Antibody Formation," *Annals of the New York Academy of Sciences* 546 (1988): pp. 175–182.
10. N. K. Jerne. "The Natural-Selection Theory of Antibody Formation," *Proceedings of the National Academy of Sciences USA* 41 (1955): pp. 849–859.
11. Joshua Lederberg to Niels Jerne, December 28, 1955, The Joshua Lederberg Papers.
12. Niels Jerne to Joshua Lederberg, 28 March 1956, The Joshua Lederberg Papers.
13. F. M. Burnet, "A Modification of Jerne's Theory of Antibody Production Using the Concept of Clonal Selection," *Australian Journal of Science* 20

(1957): pp. 67–69; Idem., F. M. Burnet, *The Clonal Selection Theory of Acquired Immunity* (Cambridge: Cambridge University Press, 1959).
14. Joshua Lederberg interview by Teicher, p. 110.
15. Ibid.
16. Ibid.
17. Burnet, "A Modification of Jerne's Theory."
18. F. M. Burnet, *Changing Patterns: An Atypical Autobiography* (Melbourne: Heinemann, 1968), p. 206.
19. Gustav Nossal interview by Jan Sapp, University of Melbourne, Melbourne, Australia, February 17, 2016.
20. Lederberg, "Ontogeny of the Clonal Selection Theory of Antibody Formation," p. 178.
21. Gustav Nossal interview.
22. G. J. V. Nossal and Joshua Lederberg, "Antibody Production by Single Cells," *Nature* 181 (1958): pp. 1419–1420.
23. J. Lederberg, "Genetic Approaches to Somatic Cell Variation: Summary Comment," *Journal of Cellular and Comparative Physiology* 52 (1958): pp. 383–401, pp. 398–400.
24. Joshua Lederberg, "Genes and Antibodies," *Science* 129 (1959): pp. 1649–1653; D. W. Talmage, "Immunological Specificity, Unique Combinations of Selected Natural Globulins Provide an Alternative to the Classical Concept," *Science* 129 (1959): pp. 1643–1648.
25. Neuberger, Michael S. "Antibody Diversification by Somatic Mutation: From Burnet Onwards," *Immunology and Cell biology* 86 (2008): pp. 124–132.
26. Lederberg, "Ontogeny of the Clonal Selection Theory of Antibody Formation," p. 180.

11 The Andromeda Man

1. Joshua Lederberg, "Our Neighbor Mars," *The Washington Post*, February 12, 1967, The Joshua Lederberg Papers, Profiles in Science, The US National Library of Medicine.
2. James Strick, "Creating a Cosmic Discipline: The Crystallization and Consolidation of Exobiology, 1957–1973," *Journal of the History of Biology* 37

(2005): pp. 131–180; Stephen Dick and James Strick, *The Living Universe: NASA and the Development of Astrobiology* (New Brunswick: Rutgers University Press, 2005).

3. Joshua Lederberg interview by Stephen Dick, West Park Hotel, Rosslyn, Virginia, November 12, 1992, p. 5., The Joshua Lederberg Papers.
4. Krishna R. Dronamraju, "On Some Aspects of the Life and Work of John Burdon Sanderson Haldane, F. R. S., in India," *Notes and Records of the Royal Society of London* 41 (1987): pp. 211–237.
5. Lederberg interview by Dick.
6. Joshua Lederberg, "Sputnik + 30," *Journal of Genetics* 66 (1987): pp. 217–220, p. 217.
7. Joshua Lederberg, "How DENDRAL was Conceived and Born," *ACM Symposium on the History of Medical Informatics*, November 5,1987, unpublished, pp. 1–6, p. 4, The Joshua Lederberg Papers.
8. A. J. Wolfe, "Germs in Space: Joshua Lederberg, Exobiology and the Public Imagination, 1958–1964," *Isis* 93 (2002): pp. 183–205.
9. Joshua Lederberg to J. B. S. Haldane, February 2, 1959, The Joshua Lederberg Papers.
10. Joshua Lederberg memo, February 24, 1958, The Joshua Lederberg Papers.
11. Joshua Lederberg, "Lunar Biology?" memo, December 1957, The Joshua Lederberg Papers.
12. Joshua Lederberg to Hugh L. Dryden, Home Secretary National Academy of Sciences, January 28, 1958; Harry Eagle to Detlev Bronk, February 6, 1958, The Joshua Lederberg Papers.
13. M. Florkin *et al.*, "Contamination by Extraterrestrial Exploration," *Nature* 183 (1959): p. 925; Idem., "Report of the Committee on the Exploration of Extraterrestrial Space (CETEX)," *ICSU Review* 1 (1959): p. 100.
14. Joshua Lederberg memo, February 24, 1958, The Joshua Lederberg Papers.
15. Joshua Lederberg to Dean Cowie, January 25, 1958, The Joshua Lederberg Papers.
16. Quoted in S. Gaylen Bradley, "Joshua Lederberg 1925–2008," *Biographical Memoirs of the National Academy of Sciences USA* 91 (2009): pp. 3–25, p. 11.
17. Joshua Lederberg to Dean Cowie, January 25, 1958, The Joshua Lederberg Papers.

18. Dean Cowie to Joshua Lederberg, February 4, 1958, The Joshua Lederberg Papers.
19. Joshua Lederberg and Dean Cowie, "Moondust", *Science* 127 (1958): pp. 1473–1475.
20. Ibid., p. 1475.
21. Ibid.
22. Ibid.
23. James Strick, "Creating a Cosmic Discipline;" Stephen Dick and James Strick, *The Living Universe*.
24. Strick, "Creating a Cosmic Discipline".
25. Ibid.
26. Lederberg used the term "exobiology" in "Report of WESTEX for the Space Science Board," October 16, 1959, The Joshua Lederberg Papers.
27. Joshua Lederberg, "Exobiology: Experimental Approaches to Life Beyond the Earth," *Science* 132 (1960): pp. 393–400.
28. Ibid., p. 397.
29. Lederberg, "Exobiology," p. 399.
30. Lederberg, "Exobiology," p. 397.
31. See also "Space Science Board Summary report of WESTEX, February 1, 1959–September 26, 1959," National Academy of Sciences and National Research Council, October 16, 1959, p. 14, The Joshua Lederberg Papers.
32. "Space Science Board Summary report of WESTEX, February 1, 1959–September 26, 1959," National Academy of Sciences and National Research Council, October 16, 1959, p. 14, The Joshua Lederberg Papers.
33. Ibid., p. 15.
34. Richard S. Young, unpublished Autobiography, May 1996, Cape Canaveral, pp. 43–44, The Joshua Lederberg Papers.
35. See Strick, "Creating a Cosmic Discipline," p. 140.
36. Joshua Lederberg to Dean Cowie, July 11, 1958, The Joshua Lederberg Papers.
37. Stanley L. Miller, "A Production of Amino Acids under Possible Primitive Earth Conditions," Science 117 (1953): pp. 528–529.
38. James Strick, "Creating a Cosmic Discipline," p. 135.
39. Harold Urey to Joshua Lederberg, November 9, 1959, The Joshua Lederberg Papers.

40. Joshua Lederberg interview by Barry Teicher, Madison Wisconsin, 1998, p. 121, The Joshua Lederberg Papers.
41. Tom Head, ed., *Conversations with Carl Sagan* (Oxford: University of Mississippi Press, 2006), p. 32.
42. Ibid.
43. Ibid.
44. See Lynn Sagan to Joshua Lederberg, January 20, 1962, The Joshua Lederberg Papers.
45. See Jan Sapp, *Evolution by Association: A History of Symbiosis* (New York: Oxford University Press, 1994); Idem., *The New Foundations of Evolution: On the Tree of Life* (New York: Oxford University Press, 2008).
46. Lederberg to Sagan, February 9, 1966, The Joshua Lederberg Papers.
47. Philip Abelson to Joshua Lederberg, January 5, 1961, The Joshua Lederberg Papers.
48. Philip Abelson, "Extra-Terrestrial Life," *Proceedings of the National Academy of Sciences* 47 (1961): pp. 575–581.
49. Joshua Lederberg to Harrison Brown, January 18, 1961, The Joshua Lederberg Papers.
50. Joshua Lederberg, "Our Neighbor Mars," *The Washington Post*, February 12, 1967, The Joshua Lederberg Papers.
51. N. H. Horowitz, R. P. Sharp, and R. W. Davies, "Planetary Contamination I: The Problem and the Agreements," *Science* 155 (1967): pp. 1501–1505; B. C. Murray, M. E. Davies, and P. K. Eckman, "Plentary Contamination II: Soviet and US Practices and Policies," *Science* 155 (1967): pp. 1505–1511; Carl Sagan, Elliott Levinthal, and Joshua Lederberg, "Contamination of Mars," *Science* 159 (1968): pp. 1191–1196.
52. Joshua Lederberg to W. A. Koshland, June 6 and June 25, 1969, The Joshua Lederberg Papers.
53. Joshua Lederberg to Robert Wise, June 1969, The Joshua Lederberg Papers.
54. Carl Sagan and Joshua Lederberg, "The Prospects for Life on Mars," *ICARUS* 28 (1976): pp. 291–300.
55. J. Mayo Greenberg, "The Structure of Evolution of Interstellar Grains," *Scientific American* (1984): pp. 124–135, p. 135.
56. Joshua Lederberg to Mayo Greenberg, July 2, 1984, The Joshua Lederberg Papers.
57. Mayo Greenberg to Joshua Lederberg, July 16, 1984, The Joshua Lederberg Papers.

12 Berkeley Debacle

1. Max Delbrück to Joshua Lederberg, February 25, 1957, The Joshua Lederberg Papers, Profiles in Science, The US National Library of Medicine.
2. R. Alexander Brink interview by Donna S. Taylor, The University of Wisconsin, University Archives Oral History Project, 1982, pp. 47–48.
3. James F. Crow, "Joshua Lederberg, 1925–2008: A Tribute," *Genetics* 178 (2008): pp. 1139–1140.
4. Eugene Lindstrom to Joshua Lederberg, September 3, 1953, The Joshua Lederberg Papers.
5. Joshua Lederberg diary entry, March 17, 1958, Filename "sand," 1956, 1958, 1960, collections of auto-notes transcribed from handwriting miscellaneous loose sheets, pp. 1–4, The Joshua Lederberg Papers, National Institutes of Health Archives, Bethesda.
6. Joshua Lederberg interview by Barry Teicher, Madison Wisconsin, June 19, 1998, transcribed by Maureen Gaber, October 6, 2008, 166pp., p. 143.
7. Joshua Lederberg to John Bowers, December 6, 1955, The Joshua Lederberg Papers.
8. Van R. Potter to E. B. Fred, February 1, 1957, The Joshua Lederberg Papers; John Bowers to President E. B. Fred, February 28, 1957, The Joshua Lederberg Papers.
9. Joshua Lederberg interview by Teicher.
10. Joshua Lederberg to Sydney Rubbo, February 28, 1957, The Joshua Lederberg Papers.
11. Joshua Lederberg to Watson, April 29, 1957, The Joshua Lederberg Papers.
12. Kenneth Thimann to Joshua Lederberg, February 28, 1957, The Joshua Lederberg Papers.
13. Ernst Mayr to Joshua Lederberg, February 28, 1957, The Joshua Lederberg Papers.
14. Joshua Lederberg to Paul Doty, March 4, 1957, The Joshua Lederberg Papers.
15. Paul Doty to Joshua Lederberg, March 1, 1957, The Joshua Lederberg Papers.
16. James Watson to Joshua Lederberg, April 3, 1957, The Joshua Lederberg Papers; James Watson to Joshua Lederberg, March 6, 1957, The Joshua Lederberg Papers; Joshua Lederberg to Thimann, April 7, 1957, The Joshua Lederberg Papers.

17. Joshua Lederberg to Anthony Iannone, August 28, 1950, The Joshua Lederberg Papers.
18. Joshua Lederberg interview by Teicher, p. 99.
19. Susan Spath, "Van Niel's Course in General Microbiology," *ASM News* 70 (2004): pp. 359–363.
20. Joshua Lederberg interview by Teicher, pp. 102–103.
21. Joshua Lederberg to Anthony Iannone, August 28, 1950, The Joshua Lederberg Papers.
22. Lederberg interview by Teicher, p. 102.
23. Joshua Lederberg and Francis Ryan to President Harland Hatcher, July 21, 1954, The Joshua Lederberg Papers.
24. Joshua Lederberg to C. A. Elvejhem, September 27, 1950, The Joshua Lederberg Papers.
25. Joshua Lederberg interview by Teicher, p. 85.
26. Joshua Lederberg to James Jenkins, January 29, 1957, The Joshua Lederberg Papers.
27. Annette Lykknes, Dolad L. Opitz, and Bridget Van Tiggelen, *For Better or For Worse? Collaborative Couples in Science*, "Epilogue," p. 274
28. Brink interview by Taylor, p. 51.
29. Joshua Lederberg to Edward Tatum, July 4, 1955, The Joshua Lederberg Papers.
30. Joshua Lederberg to James Jenkins, January 15, 1957, The Joshua Lederberg Papers.
31. Joshua Lederberg to James Jenkins, May 5, 1957; James Jenkins to Joshua Lederberg, May 14, 1957; James Jenkins to Joshua Lederberg, April 25, 1957, The Joshua Lederberg Papers.
32. James Jenkins to Joshua Lederberg, May 14, 1957, The Joshua Lederberg Papers.
33. Joshua Lederberg to Jenkins, May 16, 1957, The Joshua Lederberg Papers.
34. James Jenkins to Joshua Lederberg, January 15, 1958, The Joshua Lederberg Papers.
35. James Jenkins to Joshua Lederberg, March 13, 1958, The Joshua Lederberg Papers.
36. Joshua Lederberg to James Jenkins, March 15, 1958, The Joshua Lederberg Papers.

37. Arthur Norberg to Joshua Lederberg, May 17, 1978, The Joshua Lederberg Papers.
38. Lederberg, "Memorandum on his would be recruitment to the University of California, Berkeley in 1957," November 14, 1978, The Joshua Lederberg Papers.

13 How the West was Won

1. Avram Goldstein to William Steere, February 4, 1957, The Joshua Lederberg Papers, Profiles in Science, The US National Library of Medicine.
2. Joshua Lederberg to Arthur Kornberg, January 27, 1958, The Arthur Kornberg Papers, Profiles in Science, The US National Library of Medicine.
3. Joshua Lederberg to Edward Tatum, January 25, 1957, The Joshua Lederberg Papers.
4. Joshua Lederberg to Curt Stern, June 1, 1958, The Joshua Lederberg Papers.
5. Avram Goldstein to Joshua Lederberg, December 10, 1954, The Joshua Lederberg Papers.
6. Avram Goldstein to Joshua Lederberg, January 28, 1957, The Joshua Lederberg Papers.
7. Arthur Kornberg, *For the Love of Enzymes: The Odyssey of a Biochemist* (Cambridge: Harvard University Press, 1991).
8. Joshua Lederberg, "Fragments from a Scientific Autobiography 1969," The Joshua Lederberg Papers.
9. I. Robert Lehman, "Arthur Kornberg," *Biographical Memoirs of Fellows of the Royal Society* 58 (2012): pp. 151–161.
10. Avram Goldstein to William C. Steere, February 4, 1957, The Joshua Lederberg Papers.
11. Robert Alway to Kornberg, June 13, 1957, The Joshua Lederberg Papers.
12. Goldstein to Joshua Lederberg, June 24, 1957, The Joshua Lederberg Papers.
13. "A conversation with Henry Kaplan," Interviews by Spyros Andreopolulos conducted Dec 7, 9, and 14, 1983, p. 47, The Joshua Lederberg Papers.
14. Ibid.
15. Arthur Kornberg to Joshua Lederberg, August 19, 1957, The Joshua Lederberg Papers.

16. Joshua Lederberg to Arthur Kornberg, January 27, 1958, The Arthur Kornberg Papers.
17. Ibid.
18. Ibid.
19. Arthur Kornberg, MD, Biochemistry at Stanford, Biotechnology at DNAX, an oral history conducted by Sally Smith Hughes, Regional Oral History Office, The Bankroft Library, University of California, Berkeley, 1998.
20. Ibid.
21. Avram Goldstein to Joshua Lederberg, March 31, 1958, The Joshua Lederberg Papers.
22. Robert Alway to Joshua Lederberg, April 7, 1958, The Joshua Lederberg Papers.
23. Joshua Lederberg to Robert Alway, April 23, 1958, The Joshua Lederberg Papers.
24. Joshua Lederberg to Curt Stern, June 1, 1958, The Joshua Lederberg Papers.
25. Arthur Kornberg to Avram Goldstein, May 29, 1958, The Joshua Lederberg Papers.
26. Robert Alway to Arthur Kornberg, July 1, 1957, The Joshua Lederberg Papers.
27. Arthur Kornberg to Robert Alway, June 18, 1958, The Arthur Kornberg Papers.
28. Robert Alway to Joshua Lederberg, July 21, 1958, The Joshua Lederberg Papers.
29. Joshua Lederberg to Robert Alway, April 19, 1958, The Joshua Lederberg Papers.
30. Joshua Lederberg to A. Elvehjem, July 19, 1958, The Joshua Lederberg Papers.
31. Andreopolous, "A Conversation with Henry Kaplan," pp. 47–48.

14 Nobel Politics

1. Joshua Lederberg to Gaylen Bradley, Luca Cavalli, Phil Edwards, Tetsuo Iino, Larry Morse, Bruce Stocker, Bob Wright, and Norton Zinder, October 30, 1958, The Joshua Lederberg Papers, Profiles in Science, US National Library of Medicine.

2. Joshua Lederberg diary entry, October 26, 1958, The Joshua Lederberg Papers.
3. Sten Friberg to Joshua Lederberg, Western Union Telegram, October 30, 1958, The Joshua Lederberg Papers.
4. Erling Norrby, *Nobel Prizes and Life Sciences* (Hackensack: World Scientific Publishing, 2011), p. 191.
5. George Berry to Joshua Lederberg, July 8, 1958; Joshua Lederberg to Berry, July 15, 1958, The Joshua Lederberg Papers.
6. George Berry to Goran Liljestrand, January 25, 1958, The Joshua Lederberg Papers.
7. Ibid., p. 3.
8. George Klein, "Confronting the Holocaust: An Eyewitness Account," pp. 255–283, in Randolph L. Braham and William vanden Heuvel, eds., *The Auschwitz Reports and the Holocaust in Hungary* (New York: Columbia University Press, 2011).
9. George Klein and Eva Klein, "How one Thing has Led to Another," *Annual Reviews in Immunology* 7 (1989): pp. 1–33, p. 20.
10. Ibid.
11. Joshua Lederberg to Torbjörn Caspersson, May 13, 1955, The Joshua Lederberg Papers.
12. Joshua Lederberg to George Klein, March 17, 1958, The Joshua Lederberg Papers.
13. Norrby, *Nobel Prizes and Life Sciences*, p. 88.
14. George Klein to Joshua Lederberg, September 23, 1958; George Klein to Joshua Lederberg, October 20, 1958, The Joshua Lederberg Papers.
15. Joshua Lederberg to George Klein, October 29, 1958, The Joshua Lederberg Papers.
16. George Klein to Joshua Lederberg, November 3, 1958, The Joshua Lederberg Papers.
17. George Klein to Joshua Lederberg, November 3, 1958, The Joshua Lederberg Papers.
18. Eva Klein to Joshua Lederberg, November 20, 1958, The Joshua Lederberg Papers.
19. Quoted in Lucile Preuss, "Teamwork Pays in Bacterial Research," *The Milwaukee*, May 18, 1956, p. 4, The Joshua Lederberg Papers.

20. S. Galen Bradley to Esther Lederberg, November 10, 1958, The Joshua Lederberg Papers.
21. George P. Berry to Joshua Lederberg, December 15, 1958, The Joshua Lederberg Papers.
22. Bernard Davis to Joshua Lederberg, November 3, 1958, The Joshua Lederberg Papers.
23. H. J. Muller to Joshua Lederberg, Western Union telegraph, undated, The Joshua Lederberg Papers.
24. Edward Tatum to Joshua Lederberg, November 11, 1958, The Joshua Lederberg Papers.
25. George Beadle to Joshua Lederberg, December 30, 1958, The Joshua Lederberg Papers.
26. Joshua Lederberg to Messrs. Lisanti and Finkelstein, November 7, 1958; Jerry A. Schur to Joshua Lederberg, October 31, 1958, The Joshua Lederberg Papers.
27. Joshua Lederberg to Grayson Kirk, November 8, 1958; Grayson Kirk to Joshua Lederberg, November 20, 1958, The Joshua Lederberg Papers.
28. Francis Ryan to Joshua Lederberg, November 11, 1958, The Joshua Lederberg Papers.
29. Arthur Kornberg to Joshua Lederberg, October 31, 1958, The Joshua Lederberg Papers.
30. Wisconsin Alumnus, December 1958 issue, pp. 10–12, p. 12, The Joshua Lederberg Papers.
31. Ibid., p. 11.
32. Harriet Zuckerman and Joshua Lederberg, "Postmature Discovery?" *Nature* 324 (1986): pp. 629–631.
33. Joshua Lederberg to Karl Paul Link, November 1958, The Joshua Lederberg Papers.
34. Lederberg, "Fragments from a Scientific Autobiography 1969," 12pp., p. 4A, unpublished mss., The Joshua Lederberg Papers.
35. "Les Prix Nobel en 1958," pp. 40–41, reply to Preparation by M. E. Rudberg, quoted in Lederberg, "Fragments," p. 5, The Joshua Lederberg Papers.
36. Joshua Lederberg to George Klein, December 25, 1958, The Joshua Lederberg Papers.

37. George Beadle, "Genes and Biochemical Reactions in Neurospora," *Nobel Lecture*, December 11, 1958, pp. 587–599; Edward Tatum, "A Case History in Biological Research," *Nobel Lecture*, December 11, 1958, pp. 600–610.
38. Joshua Lederberg, "A View of Genetics," *Nobel Lecture*, May 29, 1959, pp. 615- 636, p. 615.
39. Ibid., p. 622.
40. Ibid., p. 627.
41. Ibid., p. 629.
42. Ibid., p. 624, p. 631.
43. Michael Smith, "Synthetic DNA and Biology," *Nobel Lecture*, 1993, pp. 119–136, p. 126, p. 130.

15 The New World

1. David Hamburg, "Reflections on the Career of Joshua Lederberg," in *Microbial Evolution and Co-Adaptation: A Tribute to the Life and Scientific Legacies of Joshua Lederberg: Workshop Summary* (Washington DC: National Academies Press, 2009).
2. Joshua Lederberg to Kimball Atwood, November 9, 1959, The Joshua Lederberg Papers, Profiles in Science, The US National Library of Medicine.
3. Gustav Nossal interview by Jan Sapp, Melbourne, Australia, February 16, 2016.
4. M.S. Swaminathan to Joshua Lederberg, February 8, 1958; Øjvind Winge to Joshua Lederberg, March 1958; Heinz Holter to Joshua Lederberg, May 1, 1958, The Joshua Lederberg Papers.
5. See for example A. T. Ganesan and J. Lederberg, "Physical and Biological Studies on *Bacillus subtilis* Transforming DNA." *Journal of Molecular Biology* 9 (1964): pp. 683–695; W. F Bodmer and A. T. Ganesan, "Biochemical and Genetic Studies of Integration and Recombination in *B.* subtilis Transformation," *Genetics* 50 (1964): pp. 717–738.
6. Joshua Lederberg to J. B. S Haldane, January 23, 1962, The Joshua Lederberg Papers.
7. Arthur Kornberg to Joshua Lederberg, October 23, 1958, The Joshua Lederberg Papers.

8. Joshua Lederberg to Clifford Grobstein, October 8, 1958, The Joshua Lederberg Papers.
9. See Anon, "Stanford Scientist wins Kyoto Prize for Developing Revolutionary Cell-Sorting Technology," *Stanford Medicine News Center*, June 9, 2006.
10. Walter Bodmer interview by Jan Sapp, University of Oxford, Oxford, March 20, 2017; See also Walter Bodmer and Ann Ganesan, "Joshua Lederberg," *Biographical Memoirs of Fellows of the Royal Society* 57 (2011): pp. 229–251.
11. Edward Feigenbaum interview by Jan Sapp, San Francisco, February 16, 2015.
12. Joshua Lederberg, "A Program in Genetics and Molecular Biology, Genetics Department Stanford University School of Medicine, July 1, 1965-June 30, 1968," p. 9, The Joshua Lederberg Papers.
13. Walter Bodmer interview by Jan Sapp.
14. Walter Bodmer, "Recombination and Integration in *Bacillus subtilis* Transformation: Involvement of DNA Synthesis," *Journal of Molecular Biology* 14 (1965): pp. 534–557; Walter Bodmer, "Integration of Deoxyribonuclease-Treated DNA in *Bacillis subtilis* Transformation," *Journal of General Physiology* 49 (1966): pp. 233–258; A. T. Ganesan, "Studies on In Vitro Replication of *Bacillus subtilis* DNA," *Cold Spring Harbor Symposium on Quantitative Biology* 33 (1968): pp. 45–57.
15. W. F. Bodmer, "The HLA System: Structure and Function," *Journal of Clinical Pathology* 40 (1987): pp. 948–958.
16. Walter Bodmer, "A Mathematician's Odyssey," *Annual Review of Genomics and Human Genetics* 16 (2015): pp. 1–29, p. 7.
17. R. Payne, M. Tripp, J. Weigle, W. F. Bodmer, and J. Bodmer, "A New Leukocyte Isoantigen System in Man," *Cold Spring Harbor Symposium for Quantitative Bioliogy* 29 (1964): pp. 285–295.
18. Eric Shooter interview by Thomas Ban, San Juan, Puerto Rico, December 9, 2002, The Joshua Lederberg Papers.
19. Linus Pauling *et al.*, "Sickle Cell Anemia: A Molecular Disease," *Science* 110 (1949): pp. 543–548.
20. See Vernon Ingram, "Sickle-Cell Anemia Hemoglobin: The Molecular Biology of the First 'Molecular Disease' — The Crucial Importance of Serendipity," *Genetics* 167 (2004): pp. 1–7.

21. A. B. Raper *et al.*, "Four Haemoglobins in One Individual: A Study of the Genetic Interaction of Hb-G and Hb-C," *British Medical Journal* 2 (1960): pp. 1257–1262.
22. Joshua Lederberg, "Program In Molecular Neurobiology (Submission to National Institute for Neurological Diseases and Blindness)," August 30, 1962, The Joshua Lederberg Papers.
23. Eric Shooter interview by Ban, The Joshua Lederberg Papers.
24. R. Levi-Montalcini, "The Nerve Growth Factor 35 Years Later," *Science* 237 (1987): pp. 1154–1162; Bruce Goldman, "Eric Shooter, Founding chair of Department of Neurobiology, Dies at 93," *Stanford Medicine News*, April 2018; Rita Levi-Montalcini & Pietro Calissano, "The Nerve-Growth Factor," *Scientific American* 240 (1979): pp. 44–53.
25. See Ralph Bradshaw, "Nerve Growth Factor and Related Substances: A Brief History and an Introduction to the International NGF Meeting Series," *International Journal of Molecular Sciences* 18 (2017): pp. 1143–1154.
26. Joshua Lederberg to Eric Shooter, February 27, 1964; Rita Lev-Montalcini to Joshua Lederberg, May 28, 1964, The Joshua Lederberg Papers.
27. See E. M. Shooter, "Early Days of Nerve Growth Factor Proteins," *Annual Review of Neuroscience* 24 (2001): pp. 601–629.

16 Molecular Medicine

1. Joshua Lederberg, "The Disaster of Idiocy," *The Washington Post*, October 23, 1966.
2. Clay C Whitehead to Joshua Lederberg, November 23, 1999, The Joshua Lederberg Papers, Profiles in Science, US National Library of Medicine.
3. Richard Masland to Joshua Lederberg, May 14, 1959; Joshua Lederberg to Richard Masland, June 15, 1959; Sargent Shriver to Wallace Sterling, July 31, 1959, The Joshua Lederberg Papers.
4. Joshua Lederberg, "Program in Molecular medicine and Neurobiology," Office Memorandum Draft, Stanford University, April 5, 1960, The Joshua Lederberg Papers.
5. Joshua Lederberg, "Memorandom — Research on Mental Retardation," 1962, 4pp., The Joshua Lederberg Papers.

6. Joshua Lederberg to John F. Kennedy, January 16, 1961, The Joshua Lederberg Papers.
7. "Lederberg's thoughts about Sargent Shriver," undated note, The Joshua Lederberg Papers.
8. Joshua Lederberg to R. Sargent Shriver, Jr., October 26, 1961, The Joshua Lederberg Papers.
9. Ibid.
10. Joshua Lederberg to Stanley Cohen, August 28, 1979, The Joshua Lederberg Papers.
11. Sargent Shriver to Joshua Lederberg, January 4, 1962, The Joshua Lederberg Papers.
12. Joshua Lederberg, The Joseph P. Kennedy, Jr. Laboratories for Molecular Medicine, December 1, 1961, The Joshua Lederberg Papers.
13. Joshua Lederberg, "Exobiology: Experimental Approaches to Life Beyond the Earth," *Science* 132 (1960): pp. 393–400.
14. Joshua Lederberg, "First rough draft of a chapter on the development of my work in Exobiology and in my relationships with NASA and space exploration," dictated July 15, 1986, The Joshua Lederberg Papers; Joshua Lederberg, "How DENDRAL was Conceived and Born," *ACM Symposium on the History of Medical Informatics*, National Library of Medicine, November 5, 1987, 15pp., p. 4.
15. David Perlman, "Lederberg: A Mind Exploring Man," unpublished mss., 1969, 28pp., p. 28, The Joshua Lederberg Papers.
16. Joshua Lederberg interview by Barbara Hyde, 1996, transcribed August 2002 courtesy of Pramod Srivistava, 32pp., p. 19, The Joshua Lederberg Papers.
17. Joshua Lederberg, "Cytochemical Studies of Planetary Microorganisms Explorations in Exobiology," Status Report Covering Period, July 1, 1969 to January 1, 1970; NASA Technical Report No. IRL 1105. 20pp., p. 9., The Joshua Lederberg Papers.
18. Joshua Lederberg and Elliott Levinthal, "Cytochemical Studies of Planetary Microorganisms Explorations in Exobiology," Status Report Covering Period, July 1, 1969 to January 1, 1970, for National Aeronautics and Space Administration Grant NGR-05–020-004, The Joshua Lederberg Papers.
19. Joshua Lederberg, ACME research grant application to the NIH, April 10, 1965; Joshua Lederberg, "The ACME memorandum to Faculty, Staff, and

Students interested in computer support," Stanford School of Medicine, February 14, 1967, The Joshua Lederberg Papers.

20. Joshua Lederberg, "A Program in Genetics and Molecular Biology, Genetics Department Stanford University School of Medicine, July 1, 1965-June 30, 1968," The Joshua Lederberg Papers.
21. Leonard Herzenberg *et al.*, "The History and Future of the Fluorescence Activated Cell Sorter and Flow Cytometry: A View from Stanford," *Clinical Chemistry* 48 (2002): pp. 1819–1827.
22. See Anon, "Stanford Scientist wins Kyoto Prize for Developing Revolutionary Cell-Sorting Technology," *Stanford Medicine News Center*, June 9, 2006, The Joshua Lederberg Papers.
23. M. J. Fulwyler, "Electronic Separation of Biological Cells by Volume," *Science* 150 (1965). pp. 910–911, p. 911.
24. Mario Roederer, "A Conversation with Leonard and Leonore Herzenberg," *Annual Review of Physiology* 76 (2014): pp. 1–20, p. 15
25. H. R. Hulett, W. A. Bonner, J. Barrett, and L. A. Herzenberg, "Cell Sorting: Automated Separation of Mammalian Cells as a Function of Intracellular Fluorescence," *Science* 166 (1969): pp. 747–749.
26. Joshua Lederberg, "NASA summary report," 1971, p. 46, The Joshua Lederberg Papers.

17 Breakup

1. Joshua Lederberg, "Collections of autobiographical-notes transcribed from handwriting miscellaneous loose sheets 1956–1960," Box 270, Folder 37, The National Institutes of Health Archives, National Library of Medicine, Bethesda, Maryland.
2. Joshua Lederberg, "Fragments from a Scientific Autobiography," 1969, 12pp., p. 7, The Joshua Lederberg Papers, Profiles in Science, US National Library of Medicine; Raphael Bashan, "The Jewish Heritage of the Genetics Genius," Interview with Joshua Lederberg, pp. 1–19, p. 2, November 3, 1967, Ma'arive, Israel, The Joshua Lederberg Papers.
3. Joshua Lederberg to Robert Jastrow, July 16, 1959; Herold Edward to E. L. Ginzton, January 7, 1959, The Joshua Lederberg Papers.
4. Joshua Lederberg to Edward Glinzton, April 24, 1960, The Joshua Lederberg Papers.

5. Lederberg, "Collections of Autobiographical notes."
6. Joshua Lederberg, "Autobiographical Notes re: First Wife Esther Zimmer," December 31, 1985, pp. 1–13, The Joshua Lederberg Papers.
7. Ibid., pp. 8–9.
8. Evelyn Witkins to Joshua Lederberg, August 4, 1949, The Joshua Lederberg Papers.
9. Lederberg, "Autobiographical Notes re: First Wife Esther Zimmer," pp. 10–11.
10. Ibid., pp. 8–9.
11. Lederberg, "Autobiographical Notes re: First Wife Esther Zimmer," p. 13.
12. Ibid.
13. Ibid.
14. Quoted in Alla Katsnelson, "Lederberg: A Thoughtful Visionary," *The Scientist*, February 6, 2008, The Joshua Lederberg Papers.
15. Gus Nossal interview by Jan Sapp, University of Melbourne, Australia, February 17, 2016.
16. Ibid.
17. Ibid.
18. Ann Ganesan interview by Jan Sapp, Stanford University, January 30, 2017; Walter Bodmer interview, Oxford University, March 20, 2017.
19. Clay C. Whitehead to Joshua Lederberg, November 23, 1999, The Joshua Lederberg Papers.
20. Ibid.
21. Lederberg undated diary notes, circa 1959, The Joshua Lederberg Papers, Box 270 file 30, Archives of the National Institutes of Health, The National Library of Medicine, Bethesda, Maryland, p. 28.
22. Lederberg, "Autobiographical notes re: first wife Esther Zimmer," p. 13.
23. Lederberg diary entry, June 22, 1958, Filename "sand," Box 270 file 30, Archives of the National Institutes of Health, The National Library of Medicine, Bethesda, Maryland.
24. See also Herbt Fox to Joshua Lederberg, June 17, 1958, The Joshua Lederberg Papers.
25. See also F. Aaserud, "Sputnik and the Princeton 3 — The National Security Laboratory that was Not to Be," *Studies in the Physical and Biological Sciences* 25 (1995): pp. 185–239.

26. Lederberg diary entry, June 22, 1958, Filename "sand," Box 270 file 30, The Joshua Lederberg Papers, Archives of the National Institutes of Health, Bethesda, Maryland.
27. Ibid., p. 3.
28. Ibid., p, 4.
29. Joshua Lederberg to William Nierenberg, April 23, 1984, The Joshua Lederberg Papers.
30. Joshua Lederberg to John Wheeler, April 25, 1984, The Joshua Lederberg Papers.
31. Lederberg diary notes, January 3, 1959, Collections of auto-notes transcribed from handwriting miscellaneous loose sheets, pp. 1–4, Lederberg papers, Archives of the National Institutes of Health, Bethesda, Maryland.
32. Lederberg, May 14, 1960, "Collections of auto-notes transcribed from handwriting."
33. Ibid.
34. Lederberg, August 22, 1960, "Collections of auto-notes transcribed from handwriting."
35. Ibid.
36. Lederberg, October 31, 1959, "Collections of auto-notes transcribed from handwriting miscellaneous loose sheets."
37. Joshua Lederberg to Jack Edman, December 24, 1962, The Joshua Lederberg Papers.
38. Lederberg diary entry, December 2, 1963, "Collections of auto-notes transcribed from handwriting," p. 31.
39. Joshua Lederberg to Tracy Sonneborn, November 2, 1966, The Joshua Lederberg Papers.
40. Katy Steinmetz, "Esther Lederberg and Her Husband were Both Trailblazing Scientists. Why have More People Heard of Him?" *Time*, April 11, 2019.
41. James Crow interview by Laura Small, Madison, Wisconsin, 1983, 39 pp., p. 36, The Joshua Lederberg Papers.
42. Steinmetz, "Esther Lederberg and Her Husband."
43. Jesse Ausubel interview by Jan Sapp, The Rockefeller University, New York, September 2, 2016.
44. Ibid.

45. See for example Margaret Rossiter, *Women Scientists in America: Struggles and Strategies to 1940* (Baltimore: Johns Hopkins University Press, 1982); Idem., *Women Scientists in America: Before Affirmative Action 1940*–1972 (Baltimore: Johns Hopkins University Press, 1998); Pnina Abir Am and Dorinda Outram, eds., *Uneasy Careers and Intimate Lives: Women in Science, 1789–1979* (New Brunswick: Rutgers University Press, 1987); Brenda Maddox, *Rosalind Franklin: The Dark Lady of DNA* (London: Harper Collins, 2003).
46. David Kirsch interview by Jan Sapp, Washington DC, April 10, 2015.

18 Teaching a Computer to Think

1. Joshua Lederberg to H. J. Muller, November 2, 1966, The Joshua Lederberg Papers, Profiles in Science, The US National Library of Medicine.
2. See B. Ephrussi, U. Leopold, J. D. Watson, and J. J. Weigle, "Terminology in Bacterial Genetics," *Nature* 171 (1953): p. 701; H. Quastler, ed., *Essays on the Use of Information Theory in Biology* (Urbana: University of Illinois Press, 1953); See also Lily Kay, *Who Wrote the Book of Life*? (Stanford: Stanford University Press, 2000).
3. Joshua Lederberg to Edward Ginzton, April 24, 1960, The Joshua Lederberg Papers.
4. Joshua Lederberg to Edward Ginzton, December 4, 1964, The Joshua Lederberg Papers.
5. President's Science Advisory Committee, 1963, *Science, Government, and Information*, Panel on Scientific Information, Report (A. Weinberg, chmn.), 52 pp., USGPO: Washington, The Joshua Lederberg Papers.
6. Eugene Garfield, "Citation Indexes for Science: A New Dimension in Documentation through Association of Ideas," *Science* 122 (1955): pp. 108–111; Joshua Lederberg interview by James Bohning, The Rockefeller University, December 9, 1992, The Joshua Lederberg Papers.
7. Joshua Lederberg to Garfield, May 9, 1959, The Joshua Lederberg Papers.
8. Garfield to Joshua Lederberg, May 21, 1959, The Joshua Lederberg Papers.
9. D. J. deSolla Price, "Networks of Scientific Papers: The Pattern of Bibliographic References Indicates the Nature of the Scientific Research Front," *Science* 149 (1965): pp. 510–515.

10. Eugene Garfield interview, Philadephia, June 28, 2016.
11. Lederberg interview by Bohning, p. 3.
12. Joshua Lederberg, "How DENDRAL was Conceived and Born," *ACM Symposium on the History of Medical Informatics*, National Library of Medicine, November 5, 1987, unpublished mss., 18pp., p. 5, The Joshua Lederberg Papers.
13. Joshua Lederberg to Marvine Minsky, June 22, 1955, The Joshua Lederberg Papers; M. Minsky, "Memoir on Inventing the Confocal Scanning Microscope," *Scanning* 10 (1988): pp. 128–138.
14. Joshua Lederberg, "How DENDRAL was Conceived and Born," *ACM Symposium on the History of Medical Informatics*, National Library of Medicine, November 5, 1987, unpublished mss., 18 pp., p. 5, The Joshua Lederberg Papers.
15. Lederberg, "How DENDRAL was Conceived and Born," p. 4.
16. Lederberg, "How DENDRAL was Conceived and Born," pp. 4–5.
17. Joshua Lederberg to Robert Sears and Robert Alway, May 24, 1963, The Joshua Lederberg Papers; See also Joshua Lederberg, "Electrical Technology and the Molecular Biologist," pp. 93–95, in Harlow Freitag, ed., *Electrical Engineering: The Second Century Begins* (New York: IEEE Press, 1984).
18. Joshua Lederberg, ACME research grant application to the NIH, April 10, 1965; Joshua Lederberg, "The ACME memorandum to Faculty, Staff, and Students interested in computer support," Stanford School of Medicine, February 14, 1967.
19. Walter Bodmer interview, University of Oxford, March 20, 2017.
20. He read George Boole's *The Laws of Thought* (1854) and Whitehead and Russell's three volume work *Principia Mathematica* (1925–1927), and he tried to follow theoretical biologist J. H. Woodger in his *The Axiomatic Method in Biology*, an effort to express what was then known of genetics and embryology in the formalisms of relational calculus; Joshua Lederberg, "How DENDRAL was Conceived and Born," *ACM Symposium on the History of Medical Informatics*, National Library of Medicine November 5, 1987, unpublished mss., 15pp., p. 3, The Joshua Lederberg Papers.
21. Bruce Buchanan interview by Arthur L. Norenberg, June 1–12, 1991, Pittsburg, 45pp., p. 9, The Joshua Lederberg Papers.

22. Lederberg, "How DENDRAL was Conceived and Born," p. 7.
23. Walter Bodmer and Ann Ganesan, "Joshua Lederberg", *Biographical Memoirs of the Fellows of the Royal Society* 57 (2011): pp. 229–251, p. 242.
24. Joshua Lederberg, "Dendral-64: A System for Computer Construction, Enumeration and Notation of Organic Molecules as Free Structures and Cyclic Graphs," Interim Report to the National Aeronautics and Space Administration, December 15, 1964; Joshua Lederberg, "Topological Mapping of Organic Molecules," *Proceedings of the National Academy of Sciences* 53(1) (1965): pp. 134–139.
25. Edward Shortliffe, "Presentation of the Morris F. Collen Award to Joshua Lederberg, PhD," *Journal of the American Medical Informatics Association* 7 (2000): pp. 326–332, p. 328.
26. Lederberg, "How DENDRAL was Conceived and Born," p. 8.
27. Bodmer and Ganesan, "Joshua Lederberg," p. 242.
28. Harold Klein *et al.*, "The Viking Mission Search for life on Mars," *Nature* 262 (1976): pp. 24–27.
29. See Peter Palmer and Thomas Limero, "Mass Spectrometry in the US Space Program: Past, Present, and Future," *Journal of the American Society for Mass Spectrometry* 12 (2001): pp. 656–675.
30. E. A Feigenbaum and A. H. Simon, "A Theory of the Serial Position Effect," *British Journal of Psychology*, 53 (1962): pp. 307–320; E. A. Feigenbaum and A. H. Simon, "EPAM-Like Models of Recognition and Learning," *Cognitive Science* 8 (1984): pp. 305–336.
31. Edward Feigenbaum and Julian Feldman, eds., *Computers and Thought* (San Francisco: McGraw Hill, 1963).
32. Edward Feigenbaum interview by Jan Sapp, San Francisco, February 16, 2015.
33. Feigenbaum interview by Jan Sapp.
34. E. A. Feigenbaum and R. W. Watson, "An Initial Problem for a Machine Induction Research Project," Stanford Artificial Intelligence Project Memo no. 30, April 5, 1965, The Joshua Lederberg Papers.
35. Carl Djerassi to Joshua Lederberg, July 11, 1960, The Joshua Lederberg Papers.
36. Feigenbaum interview by Jan Sapp.
37. Bill Snyder, "Joshua Lederberg: Advocacy of a 'New Literacy,'" *Stanford MD* (1978): pp. 3–4, p. 4; Carl Djerassi, *The Pill, Pygmy Chimps and Degas' Horse* (New York: Basic Books, 1992), pp. 141–142.

38. Joshua Lederberg, "SUMEX, Stanford University Medical Experimental Computer Resource," RR-00785 Annual Report submitted to Biotechnology Resource Program, National Institutes of Health, June 1, 1977, p. 81, The Joshua Lederberg Papers; B. G. Buchanan and E. H. Shortliffe, eds., *Rule-Based Expert Systems: The MYCIN Experiments of the Stanford Heuristic Programming Project* (Reading: Addison-Wesley, 1984).
39. Joshua Lederberg, "SUMEX, Stanford University Medical Experimental Computer Resource," RR-00785 Annual Report submitted to Biotechnology Resource Program, National Institutes of Health, June 1, 1977, p. 81, The Joshua Lederberg Papers.
40. Joshua Lederberg, "Note to Self," January 8, 1976, The Joshua Lederberg Papers.
41. Joshua Lederberg, "SUMEX, Stanford University Medical Experimental Computer Resource," RR-00785 Annual Report submitted to Biotechnology Resource Program, National Institutes of Health, June 1, 1977, p. 81, The Joshua Lederberg Papers.
42. Lederberg, "How DENDRAL was Conceived and Born," p. 16.
43. Ibid., p. 12.
44. Thomas Rindfleisch interview, Palo Alto, February 25, 2015 (Rindfleisch was director of the new facility. He bought the machine, put the machine together, got the software going, managed the technical staff, and with Lederberg and others, he helped to recruit the community of users).
45. J. C.R . Licklider, the head of ARPA's information processing office, had a strong interest in developing artificial intelligence research; See M. Mitchell Waldrop, *The Dream Machine: J. C. R. Licklider and the Revolution that Made Computing Personal* (New York: Viking Penguin, 2001).
46. Bill Snyder, "Joshua Lederberg: Advocacy of a 'New Literacy,'" p. 4.
47. Joshua Lederberg to Richard Garwin, February 25, 1979.
48. Lederberg, "How DENDRAL was Conceived and Born," p. 15.
49. Joshua Lederberg, "Digital Communications and the Conduct of Science: The New Literacy," *Proceedings of the IEEE* 66 (1978): pp. 1314–1319.
50. Lederberg, "Digital Communications," p. 1315.
51. Ibid., p. 1316.
52. Ibid., p. 1318.
53. Feigenbaum interview by Jan Sapp; See also Mark Stefik, *Internet Dreams: Archetypes, Myths, and Metaphors* (Cambridge: The MIT Press, 1996), p. 168.

54. Joshua Lederberg interview by Barbara Hyde, March 22, 1996, New York, Transcribed S/2002 courtesy of Pramod Srivistava, 32pp., p. 6, The Joshua Lederberg Papers.

19 The Crisis in Human Evolution

1. Joshua Lederberg to Jack Edman, December 24, 1962, The Joshua Lederberg Papers, Profiles in Science, US National Library of Medicine.
2. Joshua Lederberg, "Fragments for an Autobiography," 12pp., unpublished mss., 1969, The Joshua Lederberg Papers.
3. Lederberg, "Fragments for an Autobiography," p. 10.
4. Joshua Lederberg to Paul Doty, November 15, 1962, The Joshua Lederberg Papers.
5. See for example Daniel J. Kevles, *In the Name of Eugenics* (Berkeley: University of California Press, 1985).
6. Julian Huxley, "The Future of Man — Evolutionary Aspects," pp. 1–22, in Gordon Wolstenholme, ed., *Man and His Future: A CIBA Foundation Volume* (London: J. A. Churchill, 1963), p. 3; See also Paul Weindling, "Julian Huxley and the Continuity of Eugenics in Twentieth-Century Britain," *Journal of Modern European History* 10 (2012): pp. 480–499.
7. Huxley, "The Future of Man," p. 16.
8. Ibid., p. 16; See also Julian Huxley, *Essays of a Humanist* (New York: Harper and Row, 1964).
9. Joshua Lederberg to Jack Edman, December 4, 1962, The Joshua Lederberg Papers.
10. Gordon Wolstenholme, ed., *Man and His Future: A CIBA Foundation Volume* (London: J. A. Churchill, 1963), p. 288.
11. Joshua Lederberg, "Biological Future of Man," in Gordon Wolstenholme, ed., *Man and His Future: A CIBA Foundation Volume* (London: J. A. Churchill, 1963), pp. 263–273, p. 265.
12. Ibid.
13. Lederberg comments in *Man and His Future*, p. 361.
14. Lederberg, "Biological Future of Man," p. 269.
15. Ibid., p. 273.
16. Ibid., p. 268.

17. Ibid., p. 269.
18. J. Karamehic *et al.*, "Transplantation of Organs: One of the Greatest Achievements in History of Medicine," *Medical Archives* 62 (2008): pp. 307–310.
19. Lederberg comments in *Man and His Future*, p. 374.
20. Joshua Lederberg to Jack Edman, December 24, 1962, The Joshua Lederberg Papers.
21. Ibid.
22. Joshua Lederberg to Aldous Huxley, January 23, 1963, The Joshua Lederberg Papers.
23. Joshua Lederberg diary notes, April 7, 1963, The Joshua Lederberg Papers.
24. Joshua Lederberg, "Molecular Biology, Eugenics and Euphenics," *Nature* 198 (1963): pp. 428–429, p. 428.
25. Ibid., p. 429.
26. Joshua Lederberg, "A Crisis in Evolution," *New Scientist* 21 (1963): pp. 212–215, p. 212.
27. Ibid.
28. Joshua Lederberg, "Experimental Genetics and Human Evolution," *American Naturalist* 100 (1966): pp. 519–531, p. 525.
29. Joshua Lederberg, "The Genetics of Human Nature," *Social Research* 40 (1973): pp. 375–406, p. 379.
30. Ibid.
31. Ibid.
32. David Hamburg interview by Jan Sapp, Nuclear Threat Initiative, Washington DC, May 20, 2016.
33. Ibid.
34. Ibid.

20 The Communicator

1. Joshua Lederberg, "Fragments for an Autobiography," unpublished, 1969, p. 10, The Joshua Lederberg Papers, Profiles in Science, The US National Library of Medicine.
2. Joshua Lederberg to Jack Edman, December 24, 1962, The Joshua Lederberg Papers.

3. Joshua Lederberg to Howard Simons, March 16, 1961, The Joshua Lederberg Papers.
4. Joshua Lederberg to Howard Simon, July 8, 1966, The Joshua Lederberg Papers.
5. See Dronamraju Krishna, ed., *What I Require from Life: Writings on Science and Life from J. B. S. Haldane* (with a foreword by Sir Arthur C. Clarke and preface by James F. Crow) (Oxford: Oxford University Press, 2009), p. 145.
6. See Dronamraju Krishna, ed., *What I Require from Life: Writings on Science and Life from J. B. S. Haldane* (with a foreword by Sir Arthur C. Clarke and preface by James F. Crow) (Oxford: Oxford University Press, 2009), p. 145.
7. David Hamburg interview by Jan Sapp, May 20, 2016, Washington DC.
8. H. J. Muller to Lederberg, October 24, 1966, The Joshua Lederberg Papers.
9. Sidney Drell interview by Jan Sapp, February 17, 2015, Stanford University.
10. Joshua Lederberg, "Virginia 'Biology' Based on Delusion," *The Washington Post*, October 2, 1966.
11. "Firing Line with William F. Buckley Jr.: Shockley's Thesis Episode S0145," Recorded June 10, 1974, Published 27 Jan 2017; Wolfgang Saxon,"William B. Shockley, 79, Creator of Transistor and Theory on Race," *New York Times*, August 14, 1989; See also William H. Tucker, *Science, Politics and Racial Research* (Urbana: University of Illinois Press, 1996).
12. Walter Bodmer *et al.*, "The Issue of 'Bad Heredity,'" *Stanford MD* 5(2) (1966): p. 41.
13. Ibid.
14. Arthur Jensen, "How Much can We Boost IQ and Scholastic Achievement?" *Harvard Educational Review* 39 (1969): pp. 1–12.
15. See Ernst Mayr to Francis Crick, April 14, 1971; John Edsall to Francis Crick, March 5, 1971; Theodosius Dobzhansky to David Hamburg, October 2, 1966, The Joshua Lederberg Papers, Profiles in Science; Richard Lewonton, "Race and Intelligence," *Bulletin of Atomic Scientists* (1970): pp. 1–8.
16. Joshua Lederberg, "Comment on Shockley's Accusations of 'Lysenkoism' and on His Imputations of Racial Inferiority," unpublished, August 21, 1969, 3pp., p. 1.

17. Ibid., p. 3.
18. Ibid., p. 2.
19. Lederberg, "Race and Intelligence," *The Washington Post*, March 29, 1969, The Joshua Lederberg Papers.
20. Joshua Lederberg, "Biology could Add Little to Hitlerian Repertoire," *The Washington Post*, February 28, 1970.
21. Joshua Lederberg to Eunice Kennedy Shriver, April 16, 1973, The Joshua Lederberg Papers.
22. Joshua Lederberg, "Unpredictable Variety Still Rules Human Reproduction," *The Washington Post*, September 30, 1967.
23. Leon Kass, "Genetic Tampering," *The Washington Post*, October 30, 1967, The Joshua Lederberg Papers.
24. Joshua Lederberg to Leon Kass, October 9, 1967, The Joshua Lederberg Papers.
25. Joshua Lederberg, "Curbs on Human Engineering can Create Thought Control," *The Washington Post*, October 21, 1967.
26. Joshua Lederberg, "Biological Goal: Human Welfare," *The New York Times*, January 12, 1970.
27. Joshua Lederberg, "Egg Transplants: No End of the World," *The Washington Post*, June 20, 1971.
28. Joshua Lederberg, "We're So Accustomed to Using Chlorine that We Tend to Overlook its Toxicity," *The Washington Post*, May 3, 1969; Idem., "Ample Evidence for Taking Lead out of Gasoline," *The Washington Post*, January 17, 1970; Idem., "Politics Nullifies Science in Environmental Studies," *The Washington Post*, June 13, 1970.

21 The Advocate

1. Joshua Lederberg, "A Treaty Proposal on Germ Warfare," *The Washington Post*, September 24, 1966.
2. Joshua Lederberg interview by Jan Sapp, The Rockefeller University, July 26, 1994, 36pp., p. 8, The Joshua Lederberg Papers, Profiles in Science, US National Library of Medicine.
3. Joshua Lederberg, "The Control of Chemical and Biological Weapons," *The Stanford Journal of International Studies* 7 (1972): pp. 22–44.

4. See William Lanouette and Bela Silard, *Genius in the Shadows: A Biography of Leo Szilard, the Man Behind the Bomb* (Chicago: The University of Chicago Press, 1994).
5. Joshua Lederberg to Leo Szilard, February 18, 1955, The Joshua Lederberg Papers.
6. Linus Pauling to Joshua Lederberg, November 6, 1957, The Joshua Lederberg Papers.
7. Joshua Lederberg to Linus Pauling unsent reply, November 1957, The Joshua Lederberg Papers.
8. Bertrand Russell to Joshua Lederberg, March 14, 1959, The Joshua Lederberg Papers.
9. Joshua Lederberg to Joseph Rotblat, March 31, 1959, The Joshua Lederberg Papers.
10. Joshua Lederberg to Joseph Rotblat, September 4, 1959, The Joshua Lederberg Papers.
11. Joshua Lederberg to J. Rotblat, January 24, 1969, The Joshua Lederberg Papers.
12. Matthew Meselson interview by Jan Sapp, Harvard University, June 30, 2016.
13. Ibid.
14. Matthew Meselson to Joshua Lederberg, August 4, 1966, The Joshua Lederberg Papers.
15. John Edsall and Matthew Meselson to President Lyndon Johnson, September 19, 1966, The Joshua Lederberg Papers.
16. Joshua Lederberg to Matthew Meselson, August 3, 1966, The Joshua Lederberg Papers.
17. Ibid.
18. Meselson to Joshua Lederberg, September 16, 1966, The Joshua Lederberg Papers.
19. Joshua Lederberg, "A Treaty Proposal on Germ Warfare," *The Washington Post*, September 24, 1966.
20. Matthew Meselson to Joshua Lederberg, October 20, 1966, The Joshua Lederberg Papers.
21. Joshua Lederberg, Typescript of the Proceedings of the American Society for Microbiology's Northern California Branch debating Biological Warfare on November 11, 1967, Lederberg Papers, Profiles in Science.

22. Joshua Lederberg, "The Infamous Black Death May Return to Haunt Us," *The Washington Post,* August 31, 1968.
23. Joshua Lederberg, "Mankind had a Near Miss from a Mystery Pandemic," *The Washington Post*, September 7, 1968.
24. Friedrich Frischknecht, "The History of Biological Warfare," *EMBO Reports* (Suppl. 1) (2003): pp. S47–S52.
25. Walter Mondale to Joshua Lederberg, July 3, 1968, The Joshua Lederberg Papers; Joshua Lederberg to Walter Mondale, September 2, 1968; Joshua Lederberg to Matthew Meselson, September 13, 1968, The Joshua Lederberg Papers.
26. See Joshua Lederberg to Martin Kaplan, World Health Organization, April 3, 1969; Martin Kaplan, "The Efforts of WHO and Pugwash to Eliminate Chemical and Biological Weapons — A Memoir," *Bulletin of the World Health Organization* 77 (1999): pp. 149–155.
27. Joshua Lederberg, "The Control of Chemical and Biological Weapons," *The Stanford Journal of International Studies* 7 (1972): pp. 22–44, p. 27.
28. Joshua Lederberg, "The Control of Chemical and Biological Weapons," *The Stanford Journal of International Studies* 7 (1972): pp. 22–44, p. 28.
29. Meselson interview.
30. Judith Miller, Stephen Engelberg, and William Borad, *Germs, Biological Weapons and America's Secret War* (New York: Simon and Schuster, 2001)
31. Meselson interview.
32. Joshua Lederberg to A. Schou, December 11, 1969, The Joshua Lederberg Papers.
33. Joshua Lederberg, "Biological Warfare and the Extinction of Man," Statement before the Subcommittee on National Security Policy and Scientific Developments of the House Committee on Foreign Affairs, December 2, 1969, The Joshua Lederberg Papers.
34. Sidney Drell interview by Jan Sapp, February 15, 2015, Stanford University; David Hamburg interview, Nuclear Disarmament Initiative, Washington DC, May 20, 2016.
35. See Charles J. Conrad, Assembly Pro Tempore, *News Release*, March 24, 1970, State Capital Sacramento, California, The Joshua Lederberg Papers.
36. Lederberg interview, 1994.
37. Drell interview.

38. See Joshua Lederberg to Kenneth Yalowitz, April 7, 1970, The Joshua Lederberg Papers.
39. Joshua Lederberg to Salvador Luria, July 20, 1970, The Joshua Lederberg Papers.
40. Joshua Lederberg to Salvador Luria, July 20, 1970, The Joshua Lederberg Papers.
41. Joshua Lederberg, Memo on Conference of the Committee on Disarmament, Geneva, August 7, 1970, 13pp., p. 2, The Joshua Lederberg Papers.
42. Thomas Baily, *The Art of Diplomacy: The American Experience* (New York: Appleton-Century-Crofts, 1968); Johan Kaufman, *Conference Diplomacy: An Introduction* (Leyden: A. W. Sijthoff, 1970); Harold Jacobson and Eric Stein, *Diplomats, Scientists, and Politicians: The United States and the Nuclear Test Ban Negotiations* (Ann Arbor: University of Michigan Press, 1961).
43. Joshua Lederberg, Memo on Conference of the Committee on Disarmament, Geneva, August 7, 1970, 13pp., p. 2, The Joshua Lederberg Papers.
44. Joshua Lederberg, "Engineering Viruses for Health or Warfare," Statement made before the United National Committee on Disarmament, August 5, 1910; Joshua Lederberg, "Biological Warfare: A Global Threat," *American Scientist* 59 (1971): pp. 195–197.
45. Ibid., p. 3.
46. Joshua Lederberg, "Biological Warfare: A Global Threat," *American Scientist* 59 (1971): pp. 195–197.
47. Joshua Lederberg, Memo on Conference of the Committee on Disarmament, Geneva, August 7, 1970, 13 pp., p. 2.
48. Ibid., p. 5.
49. Joshua Lederberg, "The Control of Chemical and Biological Weapons," *The Stanford Journal of International Studies* 7 (1972): pp. 22–44.
50. Joshua Lederberg memo to Martin Kaplan, April 1972, The Joshua Lederberg Papers.
51. Ibid., p. 31.

22 Scooped

1. Joshua Lederberg to Martin Kaplan, September 23, 1974, The Joshua Lederberg Papers, Profiles in Science, US National Library of Medicine.

2. Doogab Yi, *The Recombinant University: Genetic Engineering and the Emergence of Stanford Biotechnology* (Chicago: University of Chicago Press, 2015); Arthur Kornberg, *For the Love of Enzymes: The Odyssey of a Biochemist* (Cambridge: Harvard University Press, 1989).
3. Walter Bodmer and A. T. Ganesan, "Biochemical and Genetic Studies of Integration and Recombination in *Bacillus subtlis* Transformation," *Genetics* 50 (1964): pp. 717–738; W. F. Bodmer, "Integration of Deoxyribonuclease-Treated DNA in *Bacillus subtilis* Transformation," *Journal of General Physiology* 9 (1966): pp. 233–258.
4. Joshua Lederberg, "A Program in Genetics and Molecular Biology," October 1968-October 1973, Grant proposal to the National Science Foundation, The Joshua Lederberg Papers; See also Susan Wright, *Molecular Politics* (Chicago: University of Chicago Press, 1994), p. 73.
5. Lederberg "A Program in Genetics and Molecular Biology," p. 14.
6. Ibid.
7. Joshua Lederberg, "Government is the Most Dangerous of Genetic Engineers," *The Washington Post*, July 19, 1970.
8. Quoted in David Perlman, "Lederberg: A Mind Exploring Man," unpublished mss., 1969, The Joshua Lederberg Papers, Profiles in Science, 28pp., pp. 6–7.
9. Mehran Goulian and Arthur Kornberg, "Enzymatic Synthesis of DNA, XXIII: Synthesis of Circular Replicative Form of Phage ϕ*X174* DNA," *Proceedings of the National Academy of Sciences* 58 (1967): pp. 1723–1730.
10. Kerry Grens, "Meet the Press, 1967," *The Scientist*, December 1, 2017.
11. A.T. Ganesan, "Studies on the In Vitro Synthesis of Transforming DNA," *Proceedings of the National Academy of Sciences* 61 (1968): pp. 1058–1065.
12. Ann Ganesan interview by Jan Sapp, Stanford University, January 30, 2017.
13. Walter Bodmer interview by Jan Sapp, Oxford University, March 20, 2017.
14. Luca Cavalli-Sforza to Joshua Lederberg, February 27, 1970, The Joshua Lederberg Papers.
15. V. Sgaramella, J. H. Van de Sande, and H. G. Korhana, "Studies on Polynucleotides: A Novel Joining Reaction Catalyzed by T4 Polynucleotidal Ligase," *Proceedings of the National Academy of Sciences* 87(1970): pp. 1468–1475; H. G. Korhana *et al.*, "Studies on Polynucleotides. 103/Total Synthesis of the Structural Gene for Alanine Transfer Ribonucleic Acid from Yeast," *Journal of Molecular Biology* 72 (1972): pp. 209–217.

16. Vittorio Sgaramella interview by Jan Sapp, University of Pavia, Italy, September 28, 2018.
17. Errol Freiberg, *Biography of Paul Berg: A Recombinant DNA Controversy Revisited* (World Scientific Publishing Co, 2014).
18. See Doojab Yi, *The Recombinant DNA University* (Chicago: University of Chicago Press, 2015).
19. S. N. Cohen, A. C. Y. Chang, H. Boyer, and R. B. Helling, "Construction of Biologically Functional Bacterial Plasmids *in vitro*," *Proceedings of the National Academy of Sciences USA* 70 (1973): pp. 3240–3304.
20. Joe Hedgpeth, Howard Goodman, and Herbert Boyer, "DNA Nucleotide Sequence Restricted by RI Endonuclease," *Proceedings of the National Academy of Sciences* 69 (1972): pp. 3448–3452; John Morrow and Paul Berg, "Cleavage of Simian Virus 40 DNA at a Unique Site by A Bacterial Restriction Enzyme," *Proceedings of the National Academy of Sciences* 69 (1972): pp. 3365–3369; Janet Mertz and Ronald Davis, "Cleavage of DNA by R1 Restriction Endonuclease Generates Cohesive Ends," *Proceedings of the National Academy of Sciences* 69 (1972): pp. 3370–3374; Vittorio Sgaramella, "Enzymatic Oligomerization of Bacteriophage P22 DNA and of Linear Simian Virus 40 DNA," *Proceedings of the National Academy of Sciences* 69 (1972): pp. 3389–3393.
21. Sgaramella interview.
22. See for example Paul Berg and Janet Mertz, "Personal Reflections on the Origins and Emergence of Recombinant DNA Technology," *Genetics* 184 (2010): pp. 9–17; Berg's biographer dismissed Sgaramella's contribution, see Freiberg, *Biography of Paul Berg*.
23. Stanley Cohen, "Science, Biotechnology, and Recombinant DNA: A Personal History," interviews conducted by Sally Smith Hughes in 1995, Regional Oral History Office, The Bankroft Library, University of California Berkeley, 367pp., p. 194, http://digitalassets.lib.berkeley.edu/roho/ucb/text/cohen_stanley.pdf; See also Stanley Cohen, "DNA Cloning: A Personal View After 40 Years," *Proceedings of the National Academy of Sciences* 110 (2013): pp. 11521–15529.
24. Ibid.
25. Ibid., p. 21.
26. Joshua Lederberg interview by Barbara Hyde, March 22, 1996, Transcribed in 2002 courtesy of Pramod Srivistava Oral History Interview, 32pp., p. 3, The Joshua Lederberg Papers.

27. S. N. Cohen, A. C. Y. Chang, H. Boyer, and R. B. Helling, "Construction of Biologically Functional Bacterial Plasmids in Vitro," *Proceedings of the National Academy of Sciences USA* 70 (1973): pp. 3240–3244.
28. Joshua Lederberg interview by Hyde, p. 21.

23 Prometheus Unbound?

1. Joshua Lederberg to Martin Kaplan, September 23, 1974, The Joshua Lederberg Papers, Profiles in Science, US National Library of Medicine.
2. Susan Wright, *Molecular Politics: Developing American and British Regulatory Policy for Genetic Engineering, 1972–1982* (Chicago: University of Chicago Press, 1994); John Lear, *Recombinant DNA: The Untold Story* (New York: Crown, 1978); Nicholas Wade, *The Ultimate Experiment: Man-Made Evolution* (New York: Walker, 1979); Sheldon Krimsky, *Genetic Alchemy* (Cambridge: MIT Press, 1982).
3. See for example Raymond A. Zilinskas and Burke K. Zimmerman, eds., *Gene-Splicing Wars: Reflections on the Recombinant DNA Controversy* (New York: Macmillan Publishing Company, 1986).
4. Paul Berg *et al.*, "Potential Biohazards of Recombinant DNA Molecules," *Science* 185 (1974): p. 303.
5. Joshua Lederberg to Robert Stone, August 23, 1974, The Joshua Lederberg Papers.
6. "Science had Long Known Ways to Make Subhumans," *The Washington Post*, January 11, 1969.
7. Joshua Lederberg, "Those who Fear Mind Control in Future Should Look Hard Now — at Television," *The Washington Post*, June 7, 1969.
8. Jim Shapiro *et al.*, "Isolation of Pure lac Operon DNA," *Nature* 224 (1969): pp. 768–774; Nathaniel Comfort, *The Science of Human Perfection: How Genes Became the Heart of American Medicine* (New Haven: Yale University Press, 2014), p. 219.
9. Joshua Lederberg, "Biology could Add Little to Hitlerian Repertoire," *The Washington Post*, February 28, 1970.
10. See Jon Beckwith, "Gene Expression in Bacteria and Some Concerns about the Misuse of Science," *Bacteriological Reviews* 34 (1970): pp. 222–227.
11. Joshua Lederberg to Amel Menotti, August 11, 1972, The Joshua Lederberg Papers.

12. Joshua Lederberg, "Biological Goal: Human Welfare," *The New York Times*, January 12, 1970.
13. Joshua Lederberg, "Biological Goal: Human Welfare," *The New York Times*, January 12, 1970.
14. Joshua Lederberg, "Sea-Level Canal points up need for Environmental Data," *The Washington Post*, February 1, 1969.
15. Meredith Wilson, Center for Advanced Study in the Behavioral Sciences, Center Program on Science, Technology and Society, September 9, 1971, 3pp., p. 1, The Joshua Lederberg Papers.
16. Joshua Lederberg to Robert Stone, August 23, 1974, The Joshua Lederberg Papers.
17. Joshua Lederberg to Robert Stone, August 23, 1974, The Joshua Lederberg Papers.
18. Stanley Cohen to Paul Berg, August 27, 1974, The Joshua Lederberg Papers.
19. Joshua Lederberg to Martin Kaplan, September 23, 1974, The Joshua Lederberg Papers.
20. Susan Wright, "Legitimating Genetic Engineering," *Perspectives in Biology and Medicine* 44 (2001): pp. 235–247; Paul Berg, "Meetings that Changed the World: Asilomar 1975: DNA Modification Secured," *Nature* 455 (2008): pp. 290–291.
21. Susan Wright, "Legitimating Genetic Engineering," *Perspectives in Biology and Medicine* 44 (2001): pp. 235–247; Paul Berg, "Meetings that Changed the World: Asilomar 1975: DNA Modification Secured," *Nature* 455 (2008): pp. 290–291.
22. Stanley N. Cohen, MD, "Science, Biotechnology, and Recombinant DNA: A Personal History," an oral history conducted by Sally Smith Hughes in 1995, Regional Oral History Office, The Bancroft Library, University of California, Berkeley, 2009. p. 76.
23. Joshua Lederberg, "Statement Concerning Safety Hazards Associated with Research on Recombinant DNA Molecules," February 28, 1975, The Joshua Lederberg Papers; Walter Bodmer interview by Jan Sapp, Oxford University, March 20, 2017.
24. Joshua Lederberg, "DNA Breakthrough Points Way to Therapy by Virus," *The Washington Post*, January 13, 1968.
25. Ibid., p. 3.

26. See Susan Wright, "Molecular Biology or Molecular Politics? The Production of Scientific Consensus on the Hazards of Recombinant DNA Technology," *Social Studies of Science* 16 (1986): pp. 593–620.
27. Lederberg, "Statement Concerning Safety Hazards," attached comments, March 17, 1975, in response to a *Times* story on February 28, 1975; See also Joshua Lederberg Draft Memorandum to Paul Berg, March 10, 1975, The Paul Berg Papers, Profiles in Science, The US National Library of Medicine.
28. Joshua Lederberg, "DNA Splicing: Will Fear Rob Us of its Benefits?" *PRISM*, November 1975, pp. 33–37, p. 37.
29. Ibid., p. 34.
30. Ibid., p. 34.
31. Francine Robinson Simring, "Recombinant DNA Risks and Benefits," *Science* 192 (1976): p. 940.
32. Erwin Chargaff, "On the Dangers of Genetic Meddling," *Science* 192 (1976): p. 938.
33. Liebe Cavaliere, "New Strains of Life — or Death," *New York Times Magazine*, August 22, 1976; E. Chargaff, "On the Dangers of Genetic Meddling," *Science* 192 (1976): p. 938.
34. Norton Zinder, "A Personal View of the Media's Role in the Recombinant DNA War," In R. A. Zilinskas and B. K. Zimmerman, eds., *The Gene-Splicing Wars: Reflections on the Recombinant DNA Controversy* (New York: MacMillan, 1986).
35. See Kathryn Harris *et al.*, "IBCS: A Cornerstone of Public Trust in Research," pp. 201–216, in Carole Baskin and Alan P. Zelicoff, eds., *Ensuring National Biosecurity* (London: Elsevier, 2016).
36. Joshua Lederberg to Norton Zinder, June 8, 1977, The Joshua Lederberg Papers.
37. Susan Wright, "The Social Warp of Science: Writing the History of Genetic Engineering Policy," *Science, Technology and Human Values* 18 (1993): pp. 79–101. Sheldon Krimsky, "From Asilomar to Industrial Biotechnology: Risks, Reductionism and Regulation," *Science as Culture* 14 (2005): pp. 309–323.
38. See for example H. Williams Smith, "Is it Safe to Use *Escherischia coli* K 12 in Recombinant DNA Experiments?" *The Journal of Infectious Diseases* 137 (1978): pp. 655–660.

39. Paul Berg to Donald Fredrickson, August 20, 1979, The Paul Berg Papers, Profiles in Science, The US National Library of Medicine.
40. See Donald S. Frederickson, "A History of the Recombinant DNA Guidelines in the United States," prepared for the Conference on Recombinant DNA and Genetic Experimentation, Wye Collge, Kent, UK, pp. 1–4, April 1979, Presented by W. J. Garland and reprinted in James D. Watson and John Tooze, *The DNA Story: A Documentary History of Gene Cloning* (San Francisco and Oxford: W. H. Freeman and Co., 1983), p. 398.
41. Ibid.
42. D. V. Goeddel, D. G. Kleid, F. Bolivar, H. L. Heyneker, D. G. Yansura, R. Crea, T. Hirose, A. Kraszewski, K. Itakura, and A. D. Riggs, "Expression in Escherichia coli of Chemically Synthesized Genes for Human Insulin," *Proceedings of the National Academy of Sciences USA* 76 (1979): pp. 106–110.
43. David Goeddel *et al.*, "Human Leukocyte Interferon Produced by *E. coli* is Biologically Active," *Nature* 287 (1980): pp. 411–416.
44. Paul Berg, "Meetings that Changed the World: Asilomar 1975: DNA Modification Secured," *Nature* 455 (2008): pp. 290–291.
45. Joshua Lederberg, "Pandemic as a Natural Evolutionary Phenomenon," *Social Research* 55 (1988): pp. 346–359, p. 358.

24 Unexpected Turnabout

1. Sidney Drell interview by Jan Sapp, Stanford University, February 17, 2015.
2. Joshua Lederberg, Terry Lectures, Yale University, April 1989, 11pp., p. 1, The Joshua Lederberg Papers, Profiles in Science, US National Library of Medicine.
3. John A. Moore to Joshua Lederberg, April 5, 1967, The Joshua Lederberg Papers.
4. Joshua Lederberg to John A. Moore, June 26, 1967, The Joshua Lederberg Papers.
5. Joshua Lederberg interview by Sharon Zane, March 26, 1998, The Joshua Lederberg Papers.
6. David Hamburg interview by Jan Sapp, The Nuclear Threat Initiative, Washington DC, May 20, 2016; See David Hamburg, *A Model of Prevention: Life Lessons* (London: Paradigm Publishers, 2015).

7. Joshua Lederberg, "David A. Hamburg: President-Elect of AAAS," *Science* 221 (1984): pp. 431–432.
8. Stephen Morse, "Josh Remembered," *Microbial Evolution and Co-Adaptation. A Tribute to the Life and Scientific Legacies of Joshua Lederberg* (Washington DC: National Academies Press, 2009).
9. James Darnell interview by Jan Sapp, The Rockefeller University, New York, June 24, 2015.
10. Walter Bodmer interview by Jan Sapp, The University of Oxford, March 20, 2017.
11. Ed Feigenbaum, "Tribute to Joshua Lederberg," August 17, 1978.
12. Joshua Lederberg memorandum to Stanley Cohen, "History of Funding the Facilities for the Genetics Department," August 28, 1979, The Joshua Lederberg Papers.
13. Ibid.
14. Ibid.
15. Rodney Nichols interview by Jan Sapp, New York City, May 1, 2017.
16. Darnell interview.
17. Jesse Ausubel interview by Jan Sapp, The Rockefeller University, New York, September 2, 2016.
18. Harriet Zuckerman to Joshua Lederberg, November 14, 1977, The Joshua Lederberg Papers.
19. Marguerite Lederberg interview by Jan Sapp, New York City, December 6, 2014.
20. Ibid.
21. David Kirsch interview by Jan Sapp, Washington DC, April 10, 2015.
22. Marguerite Lederberg interview.
23. Ibid.
24. Ibid.
25. Kirsch interview.
26. Marguerite Lederberg interview.

25 The Rockefeller

1. Joshua Lederberg to Arthur Norberg, March 21, 1980, The Joshua Lederberg Papers, Profiles in Science, US National Library of Medicine.

2. Israel Shenker, "Rockefeller University Hit by Storm over Tenure," *The New York Times*, September 26, 1976, p. 1 and p. 51.
3. Ibid.
4. David Hamburg interview by Jan Sapp, Nuclear Threat Initiative, Washington DC, May 20, 2016.
5. David Hamburg interview.
6. Joshua Lederberg interview by Anon, p. 18.
7. Mary Jane Zimmerman interview by Jan Sapp, Cortlandt Manor, New York, October 20, 2018.
8. Personal Communication.
9. Interview with Joshua Lederberg by Anon, p. 51.
10. Ibid., p. 40.
11. Ibid., p. 56.
12. Ibid., p. 32.
13. Ibid., p. 30.
14. Maren Imhoff interview by Jan Sapp, The Rockefeller University, New York, December 5, 2017.
15. Ibid.
16. Joshua Lederberg interview by Anon, p. 27.
17. Ibid., p. 28.
18. Quoted in Alla Katsnelson, "Lederberg: A Thoughtful Visionary," *The Scientist,* February 6, 2008.
19. Jesse Ausubel interview, The Rockefeller University, New York, September 2, 2016.
20. Cynthia Greenleaf interview by Jan Sapp, Fairfield Connecticut, July 10, 2019.
21. James Darnell interview by Jan Sapp, Rockefeller University, June 24, 2015; Jesse Ausubel interview by Jan Sapp, The Rockefeller University, September 2, 2016.
22. Rod Nichols interview by Jan Sapp, The Harvard Club, New York, December 5, 2014.
23. Joshua Lederberg to Maurice Green, January 31, 1969, The Joshua Lederberg Papers.
24. David Hamburg interview by Jan Sapp.
25. Joshua Lederberg to Detlev Bronk, February 5, 1958; Frederick Seitz to Joshua Lederberg, October 10, 1962, The Joshua Lederberg Papers.

26 Advice to Presidents

1. David Hamburg interview by Jan Sapp, The Nuclear Threat Initiative, Washington DC, May 20, 2016.
2. Joshua Lederberg to Martin Kaplan, February 23, 1976, The Joshua Lederberg Papers, Profiles in Science, US National Library of Medicine.
3. Joshua Lederberg to John Gibbons, February 26, 1986, The Joshua Lederberg Papers.
4. John Gibbons to Joshua Lederberg, April 2, 1986, The Joshua Lederberg Papers.
5. See Gregg Herken, *Cardinal Choices: Presidential Science Advising from the Atomic Bomb to SDI* (New York: Oxford University Press, 1992).
6. Ibid.
7. Richard Garwin interview by Jan Sapp, Scarsdale, New York, April 19, 2016.
8. Joshua Lederberg, "Big Decisions on Big Boom," *The Washington Post*, December 25, 1969.
9. Joshua Lederberg interview by Sharon Zane, New York, March 26, 1998, 27pp., p. 5.
10. See Stephen Morse, "Joshua Lederberg Remembered," in *Microbial Evolution and Co-Adaptation: A Tribute to the Life and Scientific Legacies of Joshua Lederberg: Workshop Summary* (Washington: National Academies Press, 2009), pp. 80–88.
11. Phillip Handler to Joshua Lederberg, November 28, 1967, The Joshua Lederberg Papers.
12. Hans Bethe and John Bardeen, "Back to Science Advisors," *The New York Times*, May 17, 1986.
13. Hans Bethe to Joshua Lederberg, February 20, 1981, The Joshua Lederberg Papers.
14. Joshua Lederberg to Hans Bethe, March 1, 1981, The Joshua Lederberg Papers.
15. Joshua Lederberg, "A President Needs Discreet Science Advisers," *The New York Times*, June 1, 1986.
16. Joshua Lederberg to David Hamburg, November 5, 1986, The Joshua Lederberg Papers.
17. See David Robinson, *Carnegie Commission on Science, Technology, and Government: A Midpoint Report to the Carnegie Corporation Board*, Octo-

ber 25, 1990, unpublished, 38pp., The Joshua Lederberg Papers, Profiles in Science.
18. Ibid., pp. 8–9.
19. Anon, *Science, Technology, and Government for a Changing World*, Carnegie Commission on Science Technology and Government, New York, April, 1993.
20. Barbara Spector, "Experts Assess Carnegie Commission's Impact on US Science Policy," The Scientist 7(10) (1993), pp. 7–10, p. 7.
21. Joshua Lederberg interview by Sharon Zane, p. 7.
22. Barbara Spector, "Experts Assess Carnegie Commission's Impact on US Science Policy," *The Scientist* 7 (10) (1993), pp. 7–10, p. 8.
23. Ibid., p. 9.
24. Jesse Ausubel interview by Jan Sapp, The Rockefeller University, New York, September 2, 2016.
25. David Hamburg interview.
26. Joshua Lederberg and William Golden to Newt Gingrich, February 24, 1995, The Joshua Lederberg Papers.
27. Roger Herdman to Joshua Lederberg, July 31, 1995, The Joshua Lederberg Papers.
28. Joshua Lederberg interview by Sharon Zane, p. 6.

27 Soviet Secrets

1. Joshua Lederberg, "Observations from Lederberg During his Presence at the Joint US-USSR CISAC Meeting in Moscow in June 1985," June 12, 1985, 4pp., The Joshua Lederberg Papers, Profiles in Science, The US National Library of Medicine.
2. William Perry interview by Jan Sapp, The Hoover Institute, Stanford University, February 3, 2017.
3. Ibid.
4. Barbara Slavin and Milt Freudenheim, "The World; Haig Implied Soviet Role in Poison Warfare," *The New York Times*, September 20, 1981, p. E3; See also Julian Robinson, Jeanne Guillemin, and Matthew Meselson, "Yellow Rain in Southeast Asia: The Story Collapses," in Susan Wright, ed., *Preventing a Biological Arms Race* (Cambridge: The MIT Press, 1990), p. 220–

238; Jonathan Tucker, "The 'Yellow Rain' Controversy: Lessons for Arms Control Compliance," *The Non Proliferation Review* (2001), pp. 25–42.

5. See Judith Miller, Stephen Engelberg, and William Broad, *Germs: Biological Weapons and America's Secret War* (New York: Simon and Schuster, 2001), p. 82.
6. Bernard Nossiter, "UN Team in Doubt on Yellow Rain," *The New York Times*, November 24, 1981, p. A7.
7. Joshua Lederberg to Fred Seitz, September 15, 1981; Matthew Meselson to Joshua Lederberg, February 19, 1982, The Joshua Lederberg Papers.
8. Joshua Lederberg to Meselson, June 4, 1982, The Joshua Lederberg Papers.
9. Matthew Meselson and Julian Perry Robinson, "The Yellow Rain Affair: Lessons from a Discredited Allegation," in S. Martin, P. Levoy, and A. Clunan, eds., *Terrorism, War or Disease? Unraveling the Use of Biological Weapons* (Stanford: Stanford University Press, 2008), pp. 72–96, p. 76; Matthew Meselson, "Incident at Ban Sa Tong," memorandum, April 25, 1986, The Joshua Lederberg Papers; Anne E. Desjardins, "From Yellow Rain to Green Wheat: 25 Years of Trichothecene Biosynthesis Research," *Journal of Agriculture and Food Chemistry* 57 (2009): pp. 4478–4484; E. Marshall, "Yellow Rain Evidence Slowly Whittled Away," *Science* 233 (1986): pp. 18–19.
10. Meselsohn and Robinson, "The Yellow Rain Affair," p. 79.
11. Ibid.
12. Joshua Lederberg, "Preventing a Biological Weapons Technology Race," MS 1985, The Joshua Lederberg Papers.
13. Matthew Meselsohn, "The Biological Weapons Convention and the Sverdlovsk Anthrax Outbreak of 1979," *Journal of the Federation of American Scientists* 41(7) (1988): pp. 1–6; p. 2.
14. Vint Cerf interview by Jan Sapp, Google Headquarters, Washington DC, April 8, 2015.
15. See Joshua Lederberg to Y. A. Ovchinnikov, May 9, 1985; Joshua Lederberg, "Observations from Lederberg During his Presence at the Joint US-USSR CISAC Meeting in Moscow in June 1985," June 12, 1985, 4pp., The Joshua Lederberg Papers.
16. Milton Leitenberg and Raymond Zilinskas, *The Soviet Biological Weapons Program: A History* (Cambridge: Harvard University Press, 2012), p. 61.
17. Joshua Lederberg to A. D. Aleksandrov, November 7, 1983, The Joshua Lederberg Papers.

18. Joshua Lederberg, "Observations from Lederberg During his Presence at the Joint US-USSR CISAC Meeting in Moscow in June 1985," June 12, 1985, 4pp., p. 1, The Joshua Lederberg Papers.
19. Ibid.
20. Ibid., p. 2.
21. See Joshua Lederberg to Yuri Ovchinnikov, May 9, 1985, The Joshua Lederberg Papers.
22. Lederberg, "Observations", June 1985, p. 2.; Loren Graham, "When Ideology and Controversy Collide: The Case of Soviet Science," *The Hastings Center Report* 12 (1982): pp. 26–32, p. 26.
23. Lederberg, "Observations," p. 3.
24. Ibid., p. 4.
25. Ibid.
26. Joshua Lederberg, Summary Minutes Ad Hoc Meeting on Biological Weapons, October 18, 1985, National Academy of Sciences, Washington DC, 12pp., p. 8; December 9, 1985, The Joshua Lederberg Papers.
27. See Michael Armacost to Joshua Lederberg, August 19, 1985; Frank Press to Joshua Lederberg, August 15, 1986, The Joshua Lederberg Papers.
28. Preliminary statement by Joshua Lederberg for the Committee on International Security and Arms Control, Moscow, October 1986, The Joshua Lederberg Papers.
29. Lynn Rusten, "Meeting of the Delegation of the US National Academy of Sciences and the Academy of Sciences of the USSR on Biological Weapons," Moscow, October 8–9, 1986, The Joshua Lederberg Papers.
30. Ibid., p. 15.
31. Joshua Lederberg to Major General Philip Russell, November 20, 1986, The Joshua Lederberg Papers.
32. Philip S. Brachman to Lynn Rusten, May 29, 1987, The Joshua Lederberg Papers.
33. Lynn Rusten, "BW Meeting," September 23, 1987, Rockefeller University, 8pp., p. 3, The Joshua Lederberg Papers.
34. See Frederick Iklé to Joshua Lederberg, September 19, 1986; Joshua Lederberg to Frederick Iklé, September 5, 1987; Comments from Dr Lederberg on Discriminate Deterrence, January 15, 1988, The Joshua Lederberg Papers.
35. Joshua Lederberg to Frederick Iklé, February 8, 1988; Frederick Iklé to Joshua Lederberg, February 16, 1988, The Joshua Lederberg Papers.

36. See Milton Leitenberg and Raymond Zilinskas, *The Soviet Biological Weapons Program: A History* (Cambridge: Harvard University Press, 2012).
37. Lynn Rusten, "BW Meeting," September 23, 1987, The Rockefeller University, 8pp., p. 7; Anon, Meeting of the Delegation of the US National Academy of Sciences and the Academy of Sciences of the USSR on International Security and Arms Control, Washington, April 1–3, 1986, p. 22, The Joshua Lederberg Papers.
38. Christian Anfinsen to Joshua Lederberg, October 10, 1989, The Joshua Lederberg Papers.
39. Christian Anfinsen to Joshua Lederberg, October 10, 1989; Joshua Lederberg to Christian Anfinsen, October 19, 1989, The Joshua Lederberg Papers.
40. See Joshua Lederberg to Robert Chancock, October 23, 1989, The Joshua Lederberg Papers.
41. Joshua Lederberg to Christian Anfinsen, October 19, 1989, The Joshua Lederberg Papers.
42. Ibid.
43. See Joshua Lederberg to Robert Chanock, October 23, 1989, The Joshua Lederberg Papers.
44. Lynn Rusten, "BW Working Group Meeting Summary Minutes," US National Academy of Sciences, Washington DC, February 7, 1989; Idem., "Meeting of the Working Groups of the US National Academy of Sciences and the Academy of Sciences of the USSR on Biological Weapons Prevention," April 1–3, 1989, The Joshua Lederberg Papers.
45. David Hamburg interview by Jan Sapp, The Nuclear Threat Initiative, Washington DC, May 20, 2016.
46. Matthew Meselson interview by Jan Sapp, Harvard University, June 30, 2016.
47. Raymond A. Zilinskas, *The Soviet Biological Weapons Program and its Legacy in Today's Russia*, Center for the Study of Weapons of Mass Destruction Occasional Paper, No 11. (Washington DC: National Defense University Press, 2016), 60pp., p. 2.
48. Raymond A. Zilinskas, *The Soviet Biological Weapons Program and its Legacy in Today's Russia*, Center for the Study of Weapons of Mass Destruction Occasional Paper, No 11. (Washington DC: National Defense University Press, 2016), 60pp., p. 23.

49. M. Meselson *et al.*, "The Sverdlovsk Anthrax Outbreak of 1979," *Science* 266 (1994): pp. 1202–1208.
50. Zilinskas, *The Soviet Biological Weapons Program*, p. 2.
51. William Perry interview; Jo Husbands memorandum to Members of the CISCAC BW Working Group, July 3, 1993, 13pp., p. 2, The Joshua Lederberg Papers.
52. See Miller *et al.*, *Germs*, pp. 140–156; Susan Wright, "Terrorists and Biological Weapons: Forging the Link in the Clinton Administration," *Politics and the Life Sciences* 25 (2007): pp. 57–115, p. 67.
53. Jo Husbands memorandum to Members of the CISCAC BW Working Group, July 3, 1993, 13pp., The Joshua Lederberg Papers.

28 The Top Predator

1. Joshua Lederberg, "Crowded at the Summit: Emergent Infections and the Global Food Chain," *ASM News* 59 (1993): pp. 162–163.
2. Joshua Lederberg, "The Infamous Black Death May Return to Haunt Us," *The Washington Post*, August 31, 1968.
3. Joshua Lederberg, "Mankind had a Near Miss from a Mystery Pandemic," *The Washington Post*, September 7, 1968.
4. Joshua Lederberg, "Biological Goal: Human Welfare," *The New York Times*, January 12, 1970.
5. Infectious disease had been the leading cause of mortality in the United States at the beginning of the 20th century and accounted for 37% of deaths. But by the middle of the century, this had been diminished to 6.8% and by the late 1980s to 2.8%, see Joshua Lederberg, "Crowded at the Summit: Emergent Infections and the Global Food Chain," *ASM News* 59 (1993): pp. 162–163.
6. Lewis Thomas, *The Lives of a Cell: Notes of a Biology Watcher* (New York: Viking Press, 1974), p. 71.
7. Stephen Morse, "The Public Health Threat of Emerging Viral Disease," *American Society for Nutritional Sciences* 127 (1997): pp. 951S–957S.
8. Joshua Lederberg interview by Jan Sapp, The Rockefeller University, New York, July 26, 1995, pp. 34–35, The Joshua Lederberg Papers, Profiles in Science, US National Library of Medicine.
9. Joshua Lederberg, "Pandemic as a Natural Evolutionary Phenomenon," *Social Research* 55(3) (1988): pp. 346–359; p. 343.

10. Ibid., p 348.
11. Lederberg, "Pandemic as a Natural Evolutionary Phenomenon," p. 356.
12. Ibid., p. 358.
13. Ibid., p. 355.
14. Ibid., p. 353.
15. Ibid.
16. Joshua Lederberg, "Retrospective 'Conference of Nobel Laureates' hosted by Francois Mitterand and Elie Wiesel in January 1988," March 4, 1988, 3pp., p. 2.
17. Elie Wiesel, *Night* (New York: Bantam Books, 1960); Idem., *Dawn*, 1961; *Day*, 1962.
18. Lederberg interview by Sapp.
19. Joshua Lederberg to Nobel Committee, January 19, 1973, The Joshua Lederberg Papers.
20. Elie Wiesel to Joshua Lederberg, July 15, 1987, The Joshua Lederberg Papers.
21. Lederberg interview by Sapp.
22. Joshua Lederberg, "Retrospective 'Conference of Nobel Laureates,'" p. 1.
23. Otto Nathan, ed., *Einstein on Peace* (New York: Random House, 1988).
24. Lederberg, "Retrospective 'Conference of Nobel Laureates,'" p. 3.
25. Ibid.
26. Ibid., p. 2.
27. Ibid.
28. Ibid.
29. Ibid., p. 3.
30. Joshua Lederberg, "Medical Science, Infectious Disease and the Unity of Humankind," *Journal of the American Medical Association* 260 (1988): pp. 684–685, p. 684.
31. Ibid., p. 685.
32. Elie Wiesel to Joshua Lederberg, April 1, 1988, The Joshua Lederberg Papers.
33. Joshua Lederberg, "Ecology has All the Requisites of an Authentic Religion," *The Washington Post*, April 18, 1970.
34. See Stephen Morse, "Joshua Lederberg Remembered," pp. 74–80, in *Microbial Evolution and Co-Adaptation: A Tribute to the Life and Scien-*

tific Legacies of Joshua Lederberg: Workshop Summary (Washington DC: National Academies Press, 2009).

35. Ibid.
36. Ibid.
37. Lederberg interview by Sapp.
38. Ibid.
39. Judith Miller, Stephen Engelberg, and William Broad, *Germs: Biological Weapons and America's Secret War* (New York: Simon and Schuster, 2001), p. 90.
40. Ibid.
41. Ibid., p. 17.
42. Ibid., p. 19.
43. Morse, "Lederberg Remembered," p. 85.
44. Frederick A. Murphy, Charles H. Calisher, Robert B. Tesh, and David H. Walker, "In Memoriam: Robert Ellis Shope: 1929–2004," *Emerging Infectious Diseases* 10 (2004): pp. 762–765.
45. Lederberg interview by Sapp.
46. Lederberg, "Medical Science, Infectious Disease," p. 244.
47. Joshua Lederberg, "Viruses and Humankind," pp. 3–9, p. 8, in Stephen S. Morse, ed., *Emerging Viruses* (New York: Oxford University Press, 1993).
48. Joshua Lederberg, Robert Shope, and Stanley Oaks, eds., *Emerging Infections. Microbial Threats to Health in the United States* (Washington DC: The National Academies Press, 1992).
49. See Miller *et al.*, *Germs*, p. 194.

29 Restless Farewell

1. Joshua Lederberg to William O. Baker, October 15, 1991, The Joshua Lederberg Papers, Profiles in Science, The US National Library of Medicine.
2. James Darnell interview by Jan Sapp, The Rockefeller University, New York, June 24, 2015.
3. Maren Imhoff interview by Jan Sapp, The Rockefeller University, New York, December 5, 2017.

4. Daniel Kevles, *The Baltimore Case: A Trial of Politics, Science, and Character* (New York: W. W. Norton, 2000).
5. Joshua Lederberg, "Preliminary Report, Laboratory of Molecular Genetics and Informatics," November 2, 1991, The Joshua Lederberg Papers.
6. Zeena Nackerdien *et al.*, "Quorum Sensing Influences *Vibrio harveyi* Growth Rates in a Manner Not Fully Accounted for by the Marker Effect of Bioluminescence," *PLoS ONE*, February 27, 2008.
7. Mathew Symonds, *Softwar: An Intimate Portrait of Larry Ellison and Oracle* (New York: Simon and Schuster, 2004), p. 380.
8. Ibid., p. 384.
9. Bryan Burrough, "The Man Who Would be Gates," *Vanity Fair*, June 1997.
10. Symonds, *Softwar*, p. 384.
11. Ibid., p. 388.
12. Ibid., p. 387.
13. Joshua Lederberg, "Defense Science Board Task Force on Supercomputer Applications," Meeting minutes, June 20–21, 1983, The Joshua Lederberg Papers.
14. See "Defense Science Board Task force for Biological Defense," Meeting minutes, January 23, 1995; Joshua Lederberg to Alexia Shelokov, March 27, 1992, The Joshua Lederberg Papers.
15. Quoted in Richard Danzig, "Biological Warfare: A Nation at Risk — A Time to Act," January 1996, National Defense University Strategic Forum, Institute for National Strategic Studies, pp. 1–5, p. 5.
16. Flanagin and Lederberg "The Threat of Biological Weapons," p. 420.
17. Quoted in Richard Danzig, "Biological Warfare: A Nation at Risk — A Time to Act," January 1996, National Defense University Strategic Forum, Institute for National Strategic Studies, pp. 1–5.
18. Ibid.
19. Margaret Hamburg interview by Jan Sapp, The Nuclear Threat Initiative, Washington DC, May 20, 2016.
20. Ibid.
21. Ibid.
22. Ibid.
23. Ibid.
24. Ibid.

25. Richard Garwin interview by Jan Sapp, Scarsdale, New York, April 19, 2016.
26. Margaret Hamburg interview.
27. Ibid.
28. Ibid.
29. Annette Flanagin and Joshua Lederberg, "The Threat of Biological Weapons-Prophylaxis and Mitigation," *Journal of the American Medical Association* 276 (1996): pp. 419–420, p. 420.
30. Judith Miller, Stephen Engelberg, and William Broad, *Germs: Biological Weapons and America's Secret War* (New York: Simon and Schuster, 2001), p. 143.
31. Joshua Lederberg to Anthony Lake, June 11, 1994, The Joshua Lederberg Papers.
32. Miller *et al.*, *Germs*, pp. 236–237.
33. Ibid., p. 240.
34. Margaret Hamburg interview by Jan Sapp.
35. Joshua Lederberg, "Infectious Disease and Biological Weapons: Prophylaxis and Mitigation," *Journal of the American Medical Association* 5 (1997): pp. 435–436.
36. Miller *et al.*, *Germs*, p. 241.
37. Joshua Lederberg, "Testimony Before Senate Appropriations Committee," March 6, 1999, The Joshua Lederberg Papers.
38. Joshua Lederberg interview by Jan Sapp, New York, July 26, 1995, pp. 1–36, p. 10, The Joshua Lederberg Papers.
39. Joshua Lederberg, "Eradication of Smallpox Shouldn't End Containment," *The Washington Post*, February 21, 1970.
40. Joshua Lederberg, "Small Pox Research: Do We Retain or Burn the Vicious Bugs?" Op-ed submission to *The Washington Post*, May 14, 1999.
41. Miller *et al.*, *Germs*, p. 231.
42. Richard Danzig interview by Jan Sapp, Washington DC, June 18, 2015.
43. Ibid.
44. Ibid.
45. Joshua Lederberg, "Notes from Talk at Centennial Celebration of the Nobel Prize in Medicine and Physiology," October 20, 2001, The Joshua Lederberg Papers.
46. Joshua Lederberg, "Cell Genetics and Hereditary Symbiosis," *Physiological Reviews* 32 (1952): pp. 403–430, p. 425.
47. Joshua Lederberg, "Infectious History," *Science* 288 (2000): pp. 287–293, p. 292.

48. Ibid., p. 290.
49. Ibid., p. 292.
50. Ibid.
51. Scc Jan Sapp, *Evolution by Association: A History of Symbiosis* (New York: Oxford University Press, 1994); Idem., *The New Foundations of Evolution: On The Tree of Life* (New York: Oxford University, 2009).
52. Lederberg, "Infectious History," p. 291.
53. Ibid., p. 292.
54. Joshua Lederberg and A. T. McCray, "Ome Sweet 'Omics — A Genealogical Treasury of Words," *The Scientist* 15 (2001): p. 8.
55. See for example S. Gilbert, J. Sapp, and A. Tauber, "A Symbiotic View of Life: We have Never been Individuals," *Quarterly Review of Biology*, 87 (2012): pp. 325–341; J. Gordon, "Honor Thy Gut Symbionts Redux," *Science* 336 (2012): pp. 1251–1253.

30 An Extraordinary Life

1. Matthew Symonds, *Softwar: An intimate Portrayal of Larry Ellison and Oracle* (New York Simon and Schuster, 2004), p. 381.
2. Ibid.
3. Joshua Lederberg to Alexander Keynan, August 7, 1972, The Joshua Lederberg Papers, Profiles in Science, US National Library of Medicine.
4. David Kirsch, personal communication, March 23, 2019.
5. William Broad, "Joshua Lederberg, 82, A Nobel Winner, Dies," *The New York Times*, February 5, 2008.
6. David Hamburg, "The Life and Legacies of Joshua Lederberg," pp. 77–79, in David Relman *et al.*, eds., *Microbial Evolution and Co-Adaptation. A Tribute to the Life and Scientific Legacies of Joshua Lederberg* (Washington DC: The National Academies Press, 2009), p. 79.
7. Quoted in Stephen Morse, "Joshua Lederberg Remembered," pp. 76–94, in *Microbial Evolution and Co-Adaptation*, p. 94.
8. Nathan Comfort, *The Tangled Field* (Cambridge: Harvard University Press, 2001), p. 253.
9. Ibid.
10. Raphael Bashan, "The Jewish Heritage of the Genetics Genius," 'Ma'arive', November 3, 1967, The Joshua Lederberg Papers.

Index

Made in the USA
Middletown, DE
04 March 2025

72224083R00255